The Holy Grail of
Science

The Holy Grail of
Science

Mick Cox

Rev. date: 09/29/2020

To order additional copies of this book, contact:
Xlibris
AU TFN: 1 800 844 927 (Toll Free inside Australia)
AU Local: 0283 108 187 (+61 2 8310 8187 from outside Australia)
www.Xlibris.com.au
Orders@Xlibris.com.au
787463

CONTENTS

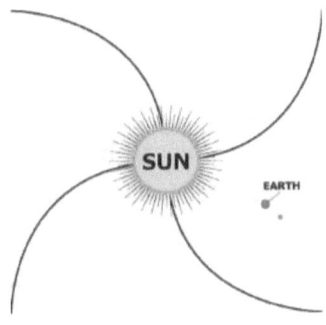

Simple Diagram of Our Solar System

There are

- four metallic planets,
- four giant gas planets,
- four large dwarf planets,
- four king tides each year,
- four pairs of neap tides each year,
- four seasons every orbit for each year,
- four years to one complete rotation of our solar system including the four lines of energy, and
- four years to every leap year.

ALL THE FOURS ABOVE ARE A DIRECT RESULT OF THE FOUR LINES. Early science recorded just **four dwarf planets,** which are the largest dwarfs, totalling eleven today, often referred to as micro-dwarfs or micro-planets.

I found this diagram which displays our sun's reactive centre mass and its four lines of magnetic energy in a gigantic science atlas in the state library twenty-six years ago. The atlas stood about 600 mm (2 ft) tall on the shelf and grabbed my attention purely due its colossal size, plus the fact that I'd never heard of a science atlas. I struggled to put it over a table, flipped it open at a random page, and saw the diagram for the first time.

In that moment, I immediately visualised our Earth orbiting around our sun, travelling on its path between the lines each year as we, here on Earth, feel the effects as the seasons change. The atlas described the diagram as our sun's magnetic energy lines, which radiate across approximately fifteen billion kilometres of our solar system to its boundaries.

It gave a further description of a **500-year cycle** called **planet alignment**, when for a brief moment all **our planets form one single ascending line of planets** every 500 years. It said that, during this time, all the planets form a single line running out into the northern hemisphere of our solar system.

It also mentioned that the next planet alignment (AD 2500) would form a line running out into the eastern hemisphere, a further turn of ninety degrees.

Let me briefly say that, since the first day in the library, I have contemplated that the planet alignment is not a perfect line of planets running out into each quadrant of our solar system but more of an alignment of their magnetic energy, meaning that the energy of each planet links up with the next planet, so the line could be considerably staggered. Also, over the last billion years since the beginning of time in our solar system, the 500-year timing has become slightly ambiguous largely due to the curve of the lines.

From what I can remember, the atlas more or less described our solar system and the line of planets behaving much like the hands of time in a solar system clock. The book stated that our solar system also rotates within four other solar systems and the magnetic lines of their sun-stars, which are identical. All these solar systems together rotate at a phenomenal rate around our galaxy; plus, our galaxy also rotates within the actual universe. I later discovered that absolutely every sun-star in our galaxy, the Milky Way, exists in this identical formation, with four lines of magnetic energy radiating from each reactive centre sphere, all 400 billion or more, with the only great difference being the size of each solar system.

I refer to our sun's reactive centre mass as merely a reactive **sphere or ball** *because the most reactive component of our sun exists on its outer skin or its crust. By referring to it as merely a sphere or ball gives the reader the impression of just an* **outer layer***. Our sun's inner core is estimated at around one million degrees Celsius whilst the crust or outer skin is estimated at a restricted six thousand degrees Celsius due to the super-sub-zero temperatures of space, and I'll tell you more about the reactive ball soon.*

Over the following months after first seeing the diagram, I considerably deliberated about the magnetic energy of these lines whilst at work in my trade as an electrical maintenance technician. I began to realise that our entire galaxy is definitely some kind of gigantic clock, with the cogs and wheels being the 400 billion solar systems, including their four magnetic lines. All are rotating together within the clock, but this book explains a great deal more indeed.

In those initial moments when peering into the diagram, I realised that as we orbit around our sun each year, here on our Earth, we feel the effects of the four seasons. I began to contemplate the dizzy spells that some people experience as we orbit through the magnetic energy lines, especially the elderly.

With its incredible facts about the 500-year cycle reeling in my mind, plus the ninety-degree rotation when realigning in AD 2500, I then lugged the giant atlas back to the shelf and set forth over the next 25 years discovering the amazing plethora of theories written in this book that seem to keep coming. Nearly every day I discover a new theory about the lines or their energy effects. This includes the rediscovery of rainbows, which I will show you soon in this book.

One of the first assimilated facts I realised is that Nostradamus, Da Vinci, Galileo, Buddha, Henry XIII, and many others of notoriety lived during or near an alignment, which may cause increased energy amongst the community, but we would discuss more on that soon. After remembering that Da Vinci sketched helicopters and submarines in the early 1500s, plus noticing that Nostradamus was one of the most amazing seers who ever lived in the 1500s, my mind turned to the presence of a strong and rather peculiar energy occurring during a planetary alignment. I began to watch the community and contemplate signs of the energy for us now and its effects during our planet alignment from that period in around 1994 until now in 2020.

After twelve years of contemplation about the diagram, one afternoon I uncovered what I identify as a 'master key' which led three straight months of sleepless nights from excitement, watching the theories unfold before my eyes, all written here in the book. This book has about 100 incredibly realistic 'theories', including how every planet was first formed on the lines, which in turn gives a very logical description of why they now exist in groups of four—four metallic planets, four giant gas planets, and four largest dwarfs—plus how our sun was formed under a 'domino's action' which formed every other sun in our galaxy, all of which I will endeavour to describe to you in a very uncomplicated, easy-to-read format. All this is really quite logical and easy to understand. The first few steps I took on this amazing path of discovery led to the most amazing but simplistic theories, written in these extremely interesting pages.

If you think the following simplest 'quick facts of fours' are incredibly daunting, wait until I tell you *how* they occurred.

The Quick Facts of Fours

In the ascending line of planets running out from the sun, the first group of four is the **four metallic planets**—Mercury, Venus, Earth, and Mars. Please note that Mercury and Mars are around half the size of Venus and our Earth.

Next in orbit, seen in an ascending line, are the **four giant gas planets**—Jupiter, Saturn, Uranus, and Neptune. Jupiter and Saturn are also much larger than Uranus and Neptune, and I'll tell you why these eight planets are in pairs by size soon.

It was early science that recorded just **four dwarf planets**, which are the largest or earliest dwarfs with a total of eleven today. Dwarf planets were once often referred to as 'micro-planets'. The first in the line of large or early dwarfs were recorded as Pluto, Haumea, Makemake, and Eris. This is the link to the ascending line of planets: http://goo.gl/images/zS1hij. If the link disappears, simply type 'planets in a line' in your search bar. The link above will initially only show the eight largest planets, but scroll down further to see the other four largest or earliest dwarfs.

When the ascending planets are in a line, there is a line of colours, which early science recorded.

Earth was blue, Mars was red, Jupiter was yellow, Saturn was orange, Uranus was amethyst, and Neptune was purple. This line of planets constitutes part of the visible light spectrum. Later in the book, I will tell you how those colours came to be formed, along with a full description of how rainbows are truly formed.

It was actually some months after discovering the diagram when I suddenly realised that every three months, when our Earth and moon orbit through a line, a **king tide** is generated. The four king tides which are generated during each yearly orbit are the largest tides every year that also occur as the seasons change.

The largest tides recorded are the 'spring tides'. The spring tide is generated twice each year because the northern hemisphere is in an opposite season to the southern hemisphere, and spring season occurs twice on Earth during each yearly orbit. Later in the book, I will describe to you what generates each spring tide, being the biggest king tide, along with the details of how a season is generated.

As previously mentioned, each group of four planets are in pairs by size; plus, the king tides are also generated in two different sizes today, which relates to the same cause for the planets in two different sizes within each group. Before and after each king tide is a neap tide (as a single tide), and therefore, there are **four pairs of neap tides during each yearly orbit**. Each neap tide is almost as high in amplitude as each king tide.

Each year when our Earth orbits between the lines, each of the **four seasons** are being generated. Some of us may sometimes also experience a peculiar energy or unbalance during the transition through the line into each season, especially the elderly folk. This often occurs whilst our Earth orbits through a diagram line, and the magnetic energy also changes in polarity.

My description of the generation of our four seasons is really quite simple. Each hemisphere of our Earth exists in a magnetic polarity of either north or south (which is actually negative and positive), and each of the four lines also exist in a magnetic polarity (north or south) which generates the weather energy for that season as we orbit between the lines. Our Earth travels between two diagram lines during each season as a part of our orbit within that particular energy of the space as we move around our sun. There is a reaction between the electrical polarities of our Earth and the electrical energy between the lines, which will suddenly change as we transition to each season. The lines in the diagram possess a polarity with a north and then another north, a south and then another south. I'll tell you a little later why the polarities during our rotation around our sun are so obscure.

The reactivity of opposite polarities and like polarities (of our Earth and the electrical energy between two lines) helps one another generate each season on our Earth in each opposite hemisphere or each opposite polarity of our Earth whilst orbiting through the energy between two lines. A north polarity on Earth above the equator and the energy generated between two lines within our solar system which might be in a south polarity help generate the specific season in the northern hemisphere, and an opposite season is generated in the southern hemisphere. There are also some other fundamental details that science describes about the tilted axis that you might already know, but as you get to know the magnetic lines throughout this book, it will very logically theorise the seasons, and I will give further details of the seasons later in the book.

Four years to every leap year - Every year our Earth orbits a further quarter of a day. Adding these ¼ days together each year equates to 125 extra days every 500 years. The extra 125 days help constitute the extra ninety degrees or quarter turn when our solar system realigns in the next quadrant. Our solar system and every other solar system in our galaxy consist of a centre spherical mass, and the four lines together rotate 125 times every 500 years, taking **four years to complete just one rotation cycle**, and I'll show you strong evidence of that fact soon.

Remember that our solar system rotates inside four other solar systems within our galaxy, there are also four rotations of each solar system during a solar system cycle, there are four planets in each group, our Earth undergoes four seasons every yearly cycle, and four king tides occur each year along with four pairs of neap tides, plus a number of other semantics in numerical fours occurring within our solar system, the galaxy, and our universe. **I conclude that our galaxy is all about 'fours'.**

Here are just a few other quick facts on our planets described in my 'slow reveals'. As mentioned, in the ascending line of planets running out from our sun, the first four are the **four metallics**; after that are the **four giant gas planets** and then the **four largest dwarf planets**, which were all discovered by early science. There are also **seven other micro-dwarfs** on record today, making a total of eleven micro-dwarfs. Science today often changes their view of the dwarfs, calling them planets one day and not planets the next. But after you've read this book, you will clearly understand why **early science identified these dwarfs as planets.**

The majority of dwarfs or micro-planets rotate randomly around their orbit throughout our solar system and never really form any part of the planet alignment. There are also actually a greater number of **unrecorded micro-planets**, up to thirty or forty of which have also been assumed to once exist in our solar system by science, which are included in my theory here today. They too were also formed in groups of fours, which I'll discuss later.

One of my many theories describes the formation of each planet, which is a very logical explanation and is the reason why each planet now exists in groups of four that matches the number of lines. I'll tell you briefly for now that

- one planet from each group of four was formed on each line,
- one metallic planet was formed on each line,
- one giant gas planet was formed on each line,
- one large dwarf planet was formed on each line.

Each of the other seven micro-dwarfs were also formed on each line, plus the thirty to forty others.

The orbital frequency of each planet in our solar system equates evenly into 500 years. This is the major factor which makes it possible for the planet alignment to occur. The orbital frequency of each planet is listed below:

Mercury - 87.97 days
Venus - 224.7 days
Earth - 365.25 days
Mars - 686.98 days
Jupiter - 4,332.82 days
Saturn - 10,755.15 days
Uranus - 30,687.15 days
Neptune - 60,190.03 days

Pluto is recorded at 248 years for every single orbit of our sun, which may be slightly inaccurate in accordance to the rules of planet alignment. This list also has the number of each planet as they ascend from our sun, and later here in this book, I will discuss why the planets increase in the number of days as they ascend out from our sun.

As mentioned, every 500 years, our entire solar system undergoes a total planetary alignment, where every planet forms one single ascending line when the energy of each aligns. The alignment takes place for just a split second in a single moment of solar system synchronicity to every planet every 500 years, but each line in the diagram identifies as a magnetic energy line and is best described in technical terms as a line of electromagnetic energy.

Soon I'll describe (in simple details) precisely what electromagnetic energy consists in. For now, I'll tell you that each electromagnetic frequency of energy consisting gamma waves through radio waves is best described on the electromagnetic spectrum (EMS) as a line of varying light frequency waves. Later, I'll give you the finer details of how some of those frequency waves are micro-sized which created micro-sized planets. This is also how the thirty to forty micro-sized planets were created, but I'll describe the details of this, plus much more, soon, and I mean much, much, much more.

For now, I'll add the following:

- Each of the four metallic planets was formed closest to our sun on each of the four lines closest to the reactive sphere. The position where each metallic was formed now reflects their position, where each metallic planet is now in formation in our solar system.
- Each of the four giant gas planets was also formed on each line after each metallic planet midway along each line. Our four giant gas planets now orbit in four positions out from our sun after the four metallic planets and now also reflect each original position where each giant gas was formed.
- Each of the four largest or earliest dwarf planets was also formed on each line in their outer positions along every line after each of the four giant gas planets.
- This also explains why the largest and earliest dwarfs are the next group of four orbiting in a random order today in the outer regions, never regularly forming a part of each planet alignment.

If you visualise each metallic planet initially formed as a magnetic energy on each line during the early years of our solar system, I can tell you that each metallic planet was initially created within a frequency which then began attracting particles specific to them. The next four giant gas planets that formed mid-way along each line existed within a frequency which was in resonance with the attracted gas particles and other matter of the same consistency which the giant gas planets are made up of today. I will explain the finer details of how these light frequencies began here soon in my slow reveals.

As described above, each planet was formed (on each line) in their ascending positions and now appears in groups as they orbit and form the line during a planet alignment. Within each group of four planets, there are also other smaller groups in pairs by size:

a pair of small metallic planets- a pair of larger metallic planets

and

a pair of small giant gas planets - a pair of larger giant gas planets

Just the eight largest planets are in the link below (as they ascend in a line away from our sun), which are clearly visible in pairs by size.

The dwarf pairs are too small to distinguish with any telescope due to the incredible distance from our Earth but are also indistinguishable due to the random orbit they possess, plus the extremely slow pace during each orbit. Pluto, for example, is recorded making a single orbit every 248 years, which may be the reason that we believe they may not ever form a part of the alignment. Go to http://goo.gl/images/zS1hij. If the link disappears, simply search 'planets in a line' and scroll down to see the other four largest and earliest dwarfs.

I'll soon describe the details of those two different types of planet groups with the groups of four, plus the cause of large and small planets, in my slow reveals which target each group of four planets, plus each group of planets in pairs by size. This book also describes the magic of each planet's colour as they appear in the ascending line which makes up just part of the visible light spectrum. In a line, they appear blue (Earth), red (Mars), yellow (Jupiter), orange (Saturn), amethyst (Uranus), and purple (Neptune). The visible light spectrum exists in the middle of the electromagnetic spectrum, suggesting that the visible light spectrum is also an important part of the EMS.

Each light frequency wave of energy on each spectrum is said to be a wave of light energy existing in each colour. The list of colours above is the earliest recorded by early science for each planet.

Today these colours seem to change as much as the dwarfs change from one science description to the next. More on these and other facts about our planets amongst these pages soon. In closing, I will add that the colours are good evidence of the planets attracting materials due to frequencies of light.

The following is a description of my theory of each king tide, plus each neap tide. According to science, every tide is generated by our sun and our moon when their gravitational energy draws on the ocean's waters. Science says that daily tides will flood to a peak every six hours and fifteen minutes and then ebb as an outgoing tide to a low tide every six hours and fifteen minutes and are generated by the sun and the moon. According to science, each of the four king tides, plus its two neap tides, is also generated by the sun and moon every three months or every quarter.

The king tide is the highest tide generated every quarter. Each neap tide is almost as high as each king tide for that period but is generated as a single tide before and after each king tide, every three months or every quarter year. Those three strongest tides, according to science, are also generated by our sun and moon, but I am about to show you my theory, which differs immensely from science. This discovery is best told in the storyline of events which unfolded to reveal my theory of the king and neap tides.

King And Neap Tides

In 1994 (just a few months after discovering the diagram) on a day at work, I suddenly realised that our Earth orbits through each line of the diagram every quarter when a king tide is generated, as well as each neap tide, which is generated on each side of a king tide. The following description is **my first slow reveal** explaining the discovery of my very first theory.

How Each King and Each Neap Tide Is Generated

Magnetic energy is described by science as an energy which is 137 times stronger than gravitational energy here on Earth. Each line of the diagram is primarily magnetic energy or electromagnetic energy. I'll give you further details describing electromagnetic energy later, but for now, all we need to know is that it is two energies (as the name implies) which function together as one.

Electrical and Magnetic Energy

In its most natural state, electromagnetic energy exists as magnetic energy and will only become identified as electrical energy once the magnetic energy has been modified within an alternator or generator but, in both circumstances, is still technically electromagnetic energy. These four magnetic lines radiate from our sun across approximately fifteen billion kilometres of our solar system to our boundaries, resulting in magnetic energy which is somewhat weaker.

Science also describes our moon's gravity at around 17% of our Earth's gravitational energy.

However, my first theory concludes that each magnetic line generates each of the three strongest tides every quarter or four times every year as our Earth and moon orbits through a magnetic diagram line, and our moon's normal gravitational energy is then amplified by the diagram line, which is far greater in magnetic energy than the gravitational energy of the moon.

Let me be clear. My first 'grail theory' says that the sun's gravitational energy is NOT an influence when each king and neap tide is generated. Each king and neap tide is generated by the energy of a diagram line when the gravitational energy of our moon is then amplified by the diagram line's magnetic energy. I'm about to show you the finer details of my first major grail theory, which explains the generation of those three strongest tides, especially the neap tides.

Each king and neap tide is generated by every magnetic line as our moon orbits through a line, and the magnetic energy of the line then amplifies our moon's normal gravitational energy when these tides are generated every quarter. The normal gravitational energy of our moon is then amplified by the magnetic energy of a diagram line which is recorded to be 137 times stronger than our Earth's gravitational energy. The amplification of the normal gravitational energy of our moon by a diagram line then generates a king tide, plus each neap tide, but this needs further explanation.

Our Earth's normal daily tides are generated by our sun and moon's general gravitational energy, and **daily tides are NOT influenced by my moon amplification theory.** *I use the term 'every quarter' because the term 'every three months' applies later in another theory.* So as to completely understand this theory on king tides, I must take you back to some of the very basic magnetic facts that we all learnt in school, plus some simple facts from my electrical trade college as an apprentice electrician, especially for the younger reader.

To completely understand my first grail theory about the three strongest tides, I need to remind you of just a few common facts of magnetics. This begins with a number of extremely basic facts taken from our early schooling days. Every magnet has a north and south pole, sometimes referred to as 'dipoles', with open-circuit lines and closed-circuit lines projecting from each pole. In our early schooling, we all saw a bar magnet placed under a piece of paper and sprinkled with iron filings, revealing the two different types of **magnetic energy lines** that exist, which are **open- and closed-circuit magnetic energy lines.** This link displays a bar magnet sprinkled with iron filings: https://www.bbc.com/bitesize/guides/z3g8d2p/revision/2. But if the link is removed, simply search for 'field display of a bar magnet in iron filings' or take a bar magnet yourself, place it under a piece of paper, and sprinkle it with iron filings.

The original solar system diagram is an extremely basic sketch which does not include every magnetic energy line. The total number of open-circuit magnetic lines projects from each **pole face of a magnet.** The original four-line diagram is an extremely simple sketch, depicting just four open-circuit lines projecting from four magnetic poles on our sun's crust, and does not include every other magnetic energy line, which will also very obviously be present. If every magnetic line was included into the diagram, it would, of course, be littered with magnetic energy lines; and of course, our solar system is actually literally littered with magnetic energy lines, which I will describe in depth throughout the book.

Each line of the diagram is a line projecting from a pole of our sun (as open-circuit lines), which also has a magnetic polarity in clockwise rotation with a north and then a south, a south and then a north. *The polarities can appear to be quite surprising in their positions— north–south and then south–north—and I'll explain the reason beyond any doubt for this with very good supporting evidence soon.*

The story behind the discovery of my first theory began, as I said, in 1994, just twenty-six years ago and just two months after seeing the diagram for the first time, when I began to wonder, 'If our moon orbits through each diagram line, is the normal gravitational energy then amplified by the magnetic energy of the solar system line, which results in a king tide?' My theory moved to 'beyond any doubt' when I looked back to my studies in electrical college. As a 16-year-old apprentice electrician, we were all taught in magnetic theories about magnetic energy lines. Let me just tell you the story of exactly what happened on that day when I suddenly looked back to what we were taught in college. It began with a conversation with a friend about tides.

After thinking for some time about my theory on king tides, I began to speak to relatives and friends about my theory involving our moon energy being amplified, which was generating each king tide. I also spoke to others around me in social circles when, on one occasion, a fellow who was a fisherman asked, 'What about the neap tides?'

I told him, 'I have no idea what a neap tide is.'

He said, 'A neap tide occurs before and after each king tide, and the amplitude of every neap tide is almost as high or as strong as each king tide of that period.'

In that exact moment, an incredibly bright light switched on in my mind as I remembered my lessons from electrical college. At absolutely every magnetic pole, there are three of the strongest lines; and during the production of electricity, these three lines alone generate 75% of all electricity during power generation.

The **centre line** from college is the strongest line, of the highest amplitude and the **greatest flux density.** Each of the other two outer lines are just slightly weaker in amplitude or magnetic energy or flux density and could possibly be referred to as secondary lines under my theories. I realised that the three strongest lines with a strong middle line and two slightly weaker outer lines are an absolute direct reflection of those three strongest or largest tides of the highest amplitude generated each time our Earth and moon orbit through the extremely simple diagram lines. The three strongest lines from college radiate from absolutely every magnetic pole, and the diagram lines very obviously radiate from each magnetic pole of our sun, out into the solar system.

Now I must point out that there are greater than three lines at every magnetic pole and that the three strongest lines exist in a line across the pole face of each magnetic pole where the generator's coil winding would pass through during every pass of each magnetic pole. But also remember that our Earth (more or less) orbits in a line across the pole face of each of our sun's four poles, but there are also a greater number of secondary lines projecting from each pole, and I'll tell you more about their magnetic field in a moment.

The king tide is the highest or strongest tide of the highest amplitude among the three largest tides. Each neap tide is almost as high or almost as strong as the king tide and occurring a number of days before and after each king tide.

The three strongest open-circuit lines which radiate from absolutely every pole face do so in a line across the front of the pole face and are a direct reflection of the three strongest tides, generated by the open-circuit lines that radiate from our sun. These lines amplify our moon's energy, but the diagram is an extremely simple sketch and does not include the other two lines or the multitude of other lines which also exist, which I will cover in a moment.

Open-circuit magnetic energy lines project from the stronger area of the pole face of every magnet. Closed-circuit magnetic energy lines project from the weaker area, which I simply identify as the pole. This link displays the iron filing experiment: https://www.bbc.com/bitesize/guides/z3g8d2p/revision/2. But if the link is removed, simply search for 'the magnetic field display of a bar magnet' or perform the simple experiment yourself. In our early schooling years, we all saw a bar magnet placed under a piece of paper and sprinkled with iron filings which displayed the magnetic energy in lines. Each group of lines was projecting from either the pole face or that weaker region, simply identified as the pole. The magnetic energy lines which project from every magnetic pole face and each magnetic pole are both open-circuit lines and closed-circuit lines, respectively.

Apart from the three strongest open-circuit lines from college, there are also other open-circuit lines which reduce in strength or flux density as they number away in formation from the centre line of the pole face. All open-circuit lines project from the pole face of every magnet, and the closed-circuit lines are more spread out, projecting from the weaker region of the remainder, which I deem as simply the pole. According to science, the exact area of the pole face is slightly ambiguous or undefined; but here in this book, I identify the pole face as the area of the magnet from where all the stronger open-circuit lines project. This includes all magnets from U-shaped magnets, electromagnets, circular magnets, horseshoe magnets, and ball-shaped magnets.

Also, according to science, the magnetic energy of every closed-circuit line flows from the south pole to the opposing north pole. When the magnetic energy of a closed-circuit line flows to the opposing north pole, the magnetic circuit is completed, and the energy is then dissipated or spent, just like a light switch when it closes, and the electricity flows. When the energy of each closed-circuit magnetic line returns to the opposing magnetic pole, the energy is then dissipated or spent and therefore identified as a closed-circuit magnetic line, thus becoming the weakest of all magnetic energy lines. *I will describe the more important or finer points of closed-circuit magnetic energy lines soon.*

However, open-circuit lines project from the pole face, travel or flow to a predetermined point out from each pole, and then simply stop, hence the term 'open-circuit magnetic lines'. And with 'no completed circuit', thus no dissipation of energy, the open-circuit lines are therefore the strongest in magnetic energy or possess the densest flux. As mentioned in our early schooling years, we all saw a bar magnet placed under the paper and sprinkled with iron filings, displaying the magnetic energy in lines by the representation of the iron filings. Each line (in the iron filings experiment) projected from either the pole face or the pole of the bar magnet, but that experiment may be very misleading when trying to understand the true form of magnetic energy lines.

That early experiment used a piece of paper to display the iron filings, which is actually only a two-dimensional view in a cross section of the magnetic energy in lines, as the paper cuts through the energy lines on a two-dimensional plane. The display on the piece of paper does not give a three-dimensional view or give an insight into the volume which exists within the energy of the magnetic energy lines. The early experiment doesn't do magnetic line energy justice. The image of the early experiment has (more or less) burnt this two-dimensional cross-sectional image into our young minds in much the same way (over a long period) that a single image on a computer screen can burn the image into the screen. The image makes magnetic lines appear to be on one flat surface, but they're very far from that form of configuration.

Each magnetic pole end (of every magnet) is identical in form to the opposite or partner pole. They can be identified as (more or less) a mirror image of each other, meaning that the north pole of each magnet is a mirror image of each south pole, with each possessing the identical number of open-circuit and closed-circuit lines. In fact, every magnetic pole is identical to every other magnetic pole in the number of magnetic lines.

Sometimes we see less magnetic lines at a particular pole end than another because one magnetic pole end is rarely stronger in flux density than the other, but each magnetic pole of every magnet has identical number of lines. Therefore, magnetic poles exist in mirrored pairs, having the same strength or flux density. The amount or level of flux also determines the length of each open-circuit line, making the lines at each pole of any magnet identical to or a mirror image of the partner and opposite pole.

Some of the finer points about magnetic lines, in lay terms, are as follows. The open-circuit magnetic lines in the early experiment displayed by the iron filings project from each pole end (of every magnet) and appear very much like the course leaves of a pineapple that protrude from the top of it. In the early childhood experiment, the piece of paper gives us the two-dimensional image as the paper cuts through the energy lines. The closed-circuit lines on the piece of paper can be seen flowing to the opposite pole reflected in iron filings (more or less) in the shape of a C. But a greater number of other closed-circuit lines also exist which are not displayed by the iron filings due to the decline in energy as the lines flow outward from pole to pole, which sometimes makes it impossible to be able to support iron filings. These other closed-circuit lines could be visualised in a C-shaped form of lines on the outer region of the magnetic field.

As I mentioned above, this image is deceiving. **The true image of these lines** is in the form of C shapes or loops but in a birdcage design with one birdcage inside another and so on. I mentioned that the piece of paper gives only a two-dimensional slice of the image through the birdcages. The open-circuit magnetic line in the centre is the only line which exists in single form with no partner line and is the strongest of all lines in flux density, which could be (more or less) visualised as the chain from which the birdcage or pineapple would be hung. This mirror image of the birdcage hanging from a centre chain line is present at both the north and south magnetic pole ends of every magnet. This includes bar magnets and electromagnets.

The early two-dimensional childhood experiment also displays a number of other open-circuit lines, plus closed-circuit lines (in C-shaped designs) outside the lines closest to the metal body of the magnet. I refer to the metal body of the magnet as the part which supports the 'major magnetic field'. The major magnetic field runs through the metallic body of every magnet.

Every magnetic line (including the major magnetic field of any electromagnet) runs through the wire coil. Each of the closed-circuit lines (in C shapes) represents the wires of the first birdcage, with a greater number of closed-circuit lines in C-shaped **birdcages** positioned outside the first or closest cage. Later, I will show you evidence of a total of seven C-shaped cages that exist for magnetic closed-circuit lines. Seven is also the total number of colours in a visual spectrum, and I'll show you how these facts relate to one another soon.

Each of these closed-circuit lines which form the C-shaped cages projects from every dipole, but as previously mentioned above, in accordance to science, the energy of every closed-circuit line projects from the south pole, travels or flows to the north pole where the magnetic energy of each line is received, and is then dissipated when the energy is spent. If you are having problems visualising this description, then it may well be due to that two-dimensional image from our early childhood having been burnt into your impressionable young mind. It might be best to read the passage above a few times until the image is replaced, although I will repeat it in a slightly differing way below.

The description above is important to be fully comprehended before reading the later chapters, where I describe the magnetic energy lines in greater depth. That early experiment (with the piece of paper and iron filings) really does do substantial damage today to the majority of our impressionable young minds when trying to consider the true shape and form of magnetic energy lines.

Just as important to again mention here is that each open-circuit line and closed-circuit line flows through the metallic body of each magnet and is identified here as the major magnetic field. From this, we will see that it is the body of the bar magnet which will be the steel structure. In the example of an electromagnet, the major magnetic field will be in the centre of the wire wound coil of the magnet.

*I refer to the body of the magnet in every example to follow as the **major magnetic field**.*

==

Personal Statement

I feel that I must make a personal statement right now to explain just why my descriptions are a little over-descriptive.

- *Mainly, I struggle a great deal with a peculiar and sometimes extreme case of dyslexia, some days not even being able to read at all, and so I need to be over-descriptive for my own benefit when re-editing.*
- *I feel it may also be important to write in an over-descriptive format so that everyone can understand, comprehend, and enjoy my theories here in the book. We have a lot further to go.*
- *I make identifications for items in distinctive fonts which are also so that I alone can re-edit the book due to my peculiar form of dyslexia.*
- *Finally, some of my writings and theories may not be comprehensible, but I assure you that I will endeavour to get a number of videos up and running because I can explain it better verbally.*
- *I have never actually read a book and most certainly have never written one. So this is a struggle, not knowing.*

Thank you for understanding.

==

Again, there are a number of magnetic closed-circuit lines flowing next to or outside of one another which make up (more or less) the formation of a number of C-shaped birdcages. Each birdcage is evenly spaced and increases in strength as they number away from the major magnetic field or the body of the magnet. Each wire of each birdcage or each magnetic line is evenly spaced from one another and also has a mirror-image partner line on the opposite side of the cage, and an even number of closed-circuit lines will therefore exist.

Remembering the piece of paper, if we were to take note of the magnetic energy lines across the pole face of every magnet, we would see open-circuit magnetic lines projecting in a straight line from the pole face across its centre. They never vary or waiver in the centre of each pole face or from one magnetic pole to the next. This makes it possible for the coil of wire to be precisely induced with electromagnetic energy during each passing of the pole face within the mechanical function of a generator during power generation (whilst the coil spins past the pole face within a power generator). *I refer to the line of open-circuit lines projecting from the centre of absolutely every dipole face as the* **major view**.

The closed-circuit magnetic lines of any magnet exist in what is best described as birdcages of lines, with the open-circuit lines projecting from the top of the birdcage (as the leaves on top of a pineapple). The centre open-circuit line (from college) also projects from the top, where the cage or pineapple might be hung from, and the mirror image of the pineapple (at the top of the bird cage) also exists at the opposite pole end. The strongest centre open-circuit line is the only single line of every magnet (including every type) and is the strongest magnetic energy line of absolutely every pole or dipole.

Later, I will tell you more important points about magnetics that also differ from what we were taught as children, which may also feel misleading. For now, we must realise the important points above that the magnetic energy displayed by the metal filings on the paper is only a two-dimensional image when the piece of paper slices through the invisible magnetic energy. It should, in fact, be visualised as a number of C-shaped birdcages in natural design, and each pole end of absolutely every magnet has the strongest single open-circuit magnetic line in the centre, projecting from where each cage or pineapple might be hung.

Also, another C-shaped birdcage of energy encompasses a larger area outside the first or primary cage, and each next cage is evenly spaced from one another, as well as each of the lines or wires of the cage in an evenly numbered, mirror-image lines. This 'next cage' or second cage is outside the primary cage, which is the closest cage to the body or major magnetic field of the magnet. There is also another cage around the second cage and another around that and so on until there are seven cages.

Each open-circuit line is evenly spaced, looking like the leaves of a pineapple, in a number of circles outside one another; plus, the wires of each circle of leaves (of open-circuit lines) are in an even number of lines with only one single line in the centre. The magnetic energy lines are identically spaced throughout the natural magnetic field in formation, but each of the seven cages around the outside of one another is also perfectly spaced.

This gives the image of a number of cages (closed-circuit lines) outside one another, but there is also the leafy end of a pineapple at each pole end of the magnet, which is the open-circuit lines. (My apologies for such a terrible description.) As the seven cages number away from the open-circuit lines, the energy of each birdcage decreases just slightly, with the strongest birdcage being closest to the body or the major magnetic field.

Ignoring for a moment the two-dimensional image given by the piece of paper as it slices through the centre of the seven birdcages, each group of the open-circuit lines appears as a circle at the top of each cage and is identical in strength or flux density and length for each magnetic pole end, but each group or circle of open-circuit lines decreases in magnetic energy as they too number away from the centre. This begins from the first circle of open-circuit lines immediately outside the centre line or the centre line as the chain from where the cage or pineapple would be hung.

The same can be said about the strength for each birdcage. Each wire of each cage or each magnetic line within each birdcage is identical in magnetic strength or flux density, but each next cage outside of that lessens just slightly in strength, with the strongest cage being closest to the major magnetic field or the metallic body of the magnet.

In the early childhood experiment, we saw a number of C-shaped lines, which were the closed-circuit lines. I described each of the C-shaped lines also forming each birdcage. This would be the image if it were possible to observe it in a three-dimensional view. You may be saying to yourself right now that there weren't seven C-shaped closed-circuit lines in that early experiment, but we must remember that the outer lines were not strong enough to support iron filings and were therefore indistinguishable by iron filings. Remember that the weakest indistinguishable outer lines (which were also in C shapes outside one another) would also be in birdcage formation.

Where the open-circuit lines project from the top of the cage is what I identify as the magnetic pole face. This is where only the open-circuit lines are projecting from, which have the strongest flux density. Most importantly, remember that absolutely every open-circuit and closed-circuit line is fixed in place, meaning that they never move or waiver, with the major view or the line of projecting open-circuit lines in a perfect line across the centre of the pole face.

The seven birdcages only consist of closed-circuit lines, with the primary birdcage possessing the strongest set of closed-circuit magnetic energy lines within any magnet, and every cage thereafter is slightly weaker in magnetic energy than the one before it. Each cage is also identically spaced apart, as well as the wires of each cage, which are evenly numbered in cage wires as the magnetic lines.

Both of the magnetic pole ends of every magnet are identical to each partner pole and therefore exist in mirrored pairs. The poles are mirrored pairs in terms of both strength of flux or flux density and the length of each magnetic line. If we look once again to the magnetic lines of any magnet, we will notice that there is a centre line in the middle of every set of open-circuit lines, and this centre line is the only single magnetic line within any magnetic field (where the cage or pineapple would be hung from).

I looked back at the diagram of our sun and realised that it only depicts four open-circuit centre lines in a two-dimensional view (as if the piece of paper had also sliced through the diagram of our solar system) as each line radiates from a magnetic pole, which is an extremely basic sketch. The simple diagram does not include any of the other open-circuit mirrored pairs of lines or closed-circuit mirrored pairs, which will also logically exist.

The lines depicted by the diagram are only the centre lines of the pole which are the strongest single lines of four different poles. (Each diagram line is where the cage or pineapple will be hung from.) Each diagram line is an open-circuit line, but if we drew a line of our Earth's orbit, it would be a line of travel much like the plain of the piece of paper; then just like the early experiment, there should be two other outer lines or the first-mirrored pair of lines, which were also included in college.

Remember that the three lines from college are a slice of paper or a two-dimensional view of the pole face as if the piece of paper had cut through this area of the pole face and the three lines are across the major view. The three lines from college were the path taken by the generator coil winding as it moved across the pole as the major view. If the two other outer lines were drawn into the diagram, they would be ever so slightly shorter and thinner, depicting an ever-so-slightly weaker pair within the major view. The major view, which includes the strongest single centre line, would appear like the view while looking down a line of trees that are planted in a pine tree planation.

If the two other outside lines from college were drawn into the diagram, it would also, of course, be easier to visualise my theory of the three strongest tides. In my amplified tide theory, the strongest centre line generates each strongest king tide, and each pair of slightly weaker lines outside the centre line generates each slightly weaker neap tide, but these two outer lines are obviously missing from the simple diagram (to state the very obvious). The three strongest magnetic lines which exist at every magnetic pole from college have an **identical 'ratio'** to the three strongest tides by strength or height and timing, plus length or duration.

My first 'holy grail' theory states that each king tide and each of the two neap tides is generated by the three strongest lines taught in college, which exist at absolutely every magnetic pole (both north and south) and very obviously also project from each pole of our sun in an identical form and ratio. The centre line from college has the highest amplitude or highest flux density when compared with each slightly weaker line on each outer side of the centre line from college in an identical ratio as each king tide compared with each slightly weaker neap tide on each of its side.

The total number of open-circuit magnetic lines projects from each **magnetic pole face.** Throughout this book, I clearly identify each pole face as the area of each pole where every open-circuit line projects from, unlike the somewhat ambiguous description given by science. Science mostly refers to the pole face as merely the flat area on top of a bar magnet, which is really undefined when considering a U-shaped magnet, a round magnet, or an electromagnet.

Flux density determines the level of magnetic strength and could be rated for magnetic potential or the potential of a magnet in its magnetic ability. The capacity or strength or flux density of an electromagnet can also be varied when changing the amount of electricity which supplies the electromagnet's coil windings. (The amount of electricity which feeds the electromagnet can be varied, which will change the magnetic strength or flux, thus flux density.) The strength or flux density of every magnet can be scientifically calculated, but each magnet should also have a rated capacity as a lay measure of its ability to perform magnetic action in a rating which might vary from 1 to 10.

The open-circuit magnetic lines has the strongest or highest flux density of every pole face of each dipole (north and south poles). Every dipole or pair of magnetic poles are also a mirrored pair of poles. Outside the pole face, each closed-circuit line also projects but are outside the open-circuit lines.

In this book, I refer to the closed-circuit lines of every magnet as the **soft energy lines** *due to being the weakest. (This is only a term used in this book.)* Each magnetic pole or every set of dipoles is identical and referred to here as **mirrored pole pairs**.

Open-circuit magnetic lines have the strongest magnetic flux, possessing the highest flux density, and every closed-circuit line thereafter is weaker. College taught us specifically about the three strongest dipole lines, meaning the three strongest lines which exist at both the north and south poles. The three strongest lines (if could be seen) are recorded in a view seen in a straight line across the pole face in much the same way that the piece of paper views the pole face.

In our early schooling, we all saw a bar magnet placed under a piece of paper and sprinkled with iron filings, revealing the two different types of **magnetic energy lines** as **open- and closed-circuit magnetic energy lines**. This link displays the iron filing experiment: https://www.bbc.com/bitesize/guides/z3g8d2p/revision/2. But if the link is removed, simply search for 'the magnetic field display of a bar magnet'.

The diagram is an extremely basic sketch which does not include every other magnetic energy line, but **according to science, the magnetic energy of closed-circuit lines** flow from the south magnetic pole to the north pole, completing a closed circuit of magnetic energy, hence the term 'closed-circuit lines'. (When a light switch is turned on, the switch is then deemed closed in electrical terms, and electrical energy then flows.)

The magnetic energy flow of every closed-circuit line is described by science as 'magnetic energy flowing from pole to pole', but my theory here refers to a slightly extended description of this flow of magnetic energy of every closed-circuit line. The energy also flows through the major magnetic field or the body of the bar magnet but begins and ends from the centre of the body of the bar magnet in a zone which I identify as the neutral magnetic zone (NMZ). You will see strong evidence of this with a complete description when I clarify the NMZ in a moment.

As I just mentioned, when the magnetic energy of every closed-circuit line returns to the opposing magnetic north pole, the energy is then dissipated. When the energy has been spent, it makes closed-circuit lines the weakest of all magnetic lines.

*It's important to understand the theory above where the metallic body of a bar magnet possesses what I refer to as the **major magnetic field**, which has the majority or highest percentage of magnetic energy flowing through it. The pole face of any magnet is the strongest area of magnetic energy or has the highest flux density, and the major magnetic field (which is the metallic body of a bar magnet) is the second strongest area. In the example of a bar magnet, it flows through the metallic body.*

Magnetic lines also flow within the major magnetic field of electromagnets, which exists within the centre of the coil of wire or the electrical winding where the iron core might be located. (More on electromagnets soon.) The magnetic energy of closed-circuit lines reduces as they number or step away from the metallic body or major magnetic field; therefore, the strongest closed-circuit lines are the two lines or cages closest to the metallic body on either side of the major magnetic field (think of a common bar magnet) and are closest to the major magnetic field of both a metallic bar magnet or any other magnet, including an electromagnet. These closed-circuit lines or cages which are closest to the body or major magnetic field of every magnet have the strongest magnetic flux or field of any closed-circuit lines. This is the first cage around the major magnetic field or the body.

Every closed-circuit magnetic line steps away thereafter from the major magnetic field in absolutely equal distances. This also applies to the major magnetic field of every other type of magnet, from U-shaped magnets to circle magnets and global or ball-shaped magnets, including electromagnets. The closed-circuit lines or cage which is the farthest from the major magnetic field or metallic body are the weakest in magnetic energy or flux of any lines. The total or exact region which is encompassed by these farthest lines is yet to be determined in its ratio of distance from the major magnetic field, but every closed-circuit line exists in what I refer to as the reducing ratio of flux density, and by the term 'reducing ratio', I mean that each next closed-circuit cage line is reduced in an identical ratio to the next line. (More on that soon.)

The distance between each next closed-circuit caged line may vary from one magnet to the next, but the distance between each closed-circuit line exists in a ratio which varies according to the flux density, and this is why the flux density should have a rating of 1–10.

The stronger or higher the flux density, the farther the distance between closed-circuit lines will be, but the distance (as stated directly above) is in a reducing ratio to flux density. This is yet to be actualised by science and is another of my holy grail theories.

I refer to the entire magnetic field of every magnet as the 'magnetic body'. Just how far from the major magnetic field the magnetic body or field extends to in this reducing ratio and the formula used to calculate the reducing ratio are yet to be discovered by science within every type of magnet, including electromagnets, but I can tell you that the distance also exists in a ratio to the physical size and flux density of the magnet, although the magnetic body or field of different types of magnets do vary.

As each closed-circuit line steps or numbers away from the major magnetic field, it reduces in flux density. The strength of each closed-circuit line (in cage formation) exists in a reducing ratio that will, one day, also be calculable when this ratio is uncovered. The reduction of the level of flux density is equal within every next set of cage lines and is a common ratio from one magnet to the next as the closed-circuit lines number away from the major magnetic field. As stated, one day it will be possible to determine the reducing ratio by using some sort of magnetic ratio formula, which is yet to be discovered by science (after they find a need for it).

The total number of open-circuit lines compared to closed-circuit lines is always the same throughout every magnet and also exists in a common ratio, meaning that the total number of open-circuit lines compared to closed-circuit lines is always the same, which is also a fixed ratio and yet to be discovered. Closed-circuit energy lines flow from the south pole to the north pole with the identical distances between each cage of lines before and after each cage as they number or step away from the major magnetic field in a formation that is actually quite concentric.

Just as for the open-circuit lines, they too would also be in concentric formation if they completed their path except for the centre line. To help you visualise the closed-circuit concentric formation of these magnetic lines, think of a stone that is thrown into a pond; the waves which are generated exist in a concentric form. The water waves are equally distanced from one another. The waves originate from one common point or source, and so too for closed-circuit magnetic lines as they number away from the source, which is the major magnetic field. *It may sound strange, but one day science will discover a relationship between the distances of closed-circuit magnetic lines and the distances between water waves in concentric circles because they are indirectly related.*

As I have stated,

- The level of flux density which is flowing within each closed-circuit cage line reduces as they number away from the major magnetic field or metallic body (in the example of a bar magnet), with the closest cage line to the major magnetic field being the strongest or highest in magnetic flux density, including electromagnets.
- The magnetic energy reduces as each closed-circuit cage line numbers or steps away from the major magnetic field in a common magnetic reducing ratio of flux density compared with the flux density total and the previous cage line.
- The distances are also identical between concentric closed-circuit lines and open-circuit lines throughout every magnetic field (just like the concentric water waves). This absolute equal or identical distance does not vary within each magnet and also exists in a fixed ratio distance determined by the level of flux and is a predetermined or fixed distance set forth by nature.

Later, science will discover the 'common reducing ratio' of magnetic energy lines and also of concentric water waves which will identify the indirect relationship between concentric magnetic lines and concentric water waves that, until now, has gone undiscovered. As far-fetched as it might seem, as you read the pages of this book, you will begin to see this relationship.

*I refer to the absolute equal or identical distances between closed-circuit magnetic lines and open-circuit magnetic lines as the **common magnetic ratio distance**. I also refer to the total number of closed-circuit lines compared with the total number of open-circuit lines as the **common magnetic ratio number**. The flux density of every closed-circuit cage line reduces in strength as they number away from one another, as they decrease in field strength or flux density in fractions which I refer to as the **common magnetic ratio density**.*

The basis of each identification above is the magnetic ratio, which will be identical throughout every formula and calculation. Each calculation will differ just slightly in distance, number, and density but will not vary in the basis of the formula, which is the magnetic ratio. When the ratio is discovered, it will also apply to the calculation of the distance and amplitude between concentric water waves. (This is how I know that they are indirectly related.)

Science describes the flow of magnetic energy within closed-circuit lines as *magnetic energy which flows from the south pole to the north pole*, but later in this book, I'll give a further detailed description, which is taken from the example of our Earth as a magnet. This will be an extended description of the magnetic flow of the closed-circuit lines, which varies slightly from the description given by science, where the magnetic energy of every closed-circuit line also flows through the major magnetic field or the body of every magnet, including electromagnets, that begins and ends in the centre.

In that detailed description, I will tell you about my theory of the flow of the closed-circuit magnetic energy lines, beginning from the middle of the major magnetic field and ending again in the middle of the major magnetic field or the metallic body (in the example of a bar magnet), which is also identical within electromagnets.

Open-Circuit Magnetic Line Energy

In regard to open-circuit magnetic line energy, science says that each open-circuit magnetic energy line simply projects from the pole face of every magnetic dipole (meaning both north and south poles), travels or flows from the pole face to a predetermined distance out from each pole, and then simply stops. This predetermined distance to which open-circuit lines flow exists in a ratio and is also dependent on the flux density of the magnet. The centre line travels or flows to the farthest distance of any open-circuit lines and then simply stops and is the strongest or highest flux density out of every magnetic line.

The distance the centre line travels to is the longest, and the flux density is the strongest compared with the distance and density of the two next outer lines which are slightly weaker out of the three strongest. At every magnetic pole face, there are three strongest open-circuit magnetic lines, but there are also other open-circuit lines which also exist that also travel and flow to a predetermined distance and then simply stop (set forth by nature). The distance that each open-circuit line travels to flows in a reducing ratio as they step away from the centre line, but each length exists in a common ratio to the next line. All other open-circuit lines also reduce in energy in my common reduction ratio as they step or number away from the previous line.

The distance that each open-circuit line travels or flows to then simply stops, plus the reducing level or amount of flux density that exists within every open-circuit line as they number or step away from the strongest centre line will one day be calculable using the common magnetic ratio formula (when it is uncovered).

The total number of open-circuit and closed-circuit lines which exist within absolutely every magnetic field is a constant, meaning the number of lines are fixed and does not vary from one magnet to the next. *The total number of open-circuit lines compared with that of closed-circuit lines differs but exists in a common ratio and is also a constant, which will one day also be calculable by comparison to find what I refer to as the* **common magnetic ratio total.** When the formulae for the common magnetic ratio distance and the common magnetic ratio number are eventually discovered, it would not surprise me if each formula is just ever so slightly different or may actually, in fact, be identical, and the source of my information is yet to be revealed.

The distance of the gap between the **open-circuit lines and closed-circuit lines** exists in the common magnetic ratio distance, which is also a constant. Each open-circuit line also travels to a specific length, which can also be calculated using the magnetic ratio distance (MRD) and is set forth by nature. (How I know this will also be revealed later.)

Again, in regard to open-circuit magnetic energy, the open-circuit lines travel or flow to a predetermined distance and then simply stop, and the distance is dependent on the flux density of the magnet. When there is no completion of a magnetic circuit, there is no dissipation, making open-circuit lines the highest in flux density of all magnetic lines.

Again, in regard to open-circuit magnetic lines,

- as apprentice electricians, we are taught in **magnetic theory** that every magnetic pole has three specific open-circuit lines projecting from the pole face, which have the strongest or highest flux density among all lines and project from absolutely every magnetic pole face;
- those three strongest lines generate 75% of the total electricity during power generation, but in college, those lines were simply referred to as open-circuit lines;
- of those three lines, the strongest is the centre line, possessing the highest flux density, and each next outer line (each side) is just slightly weaker.

The other 25% consists of the **other open-circuit lines**, plus the total number of **closed-circuit lines,** with the closed-circuit lines being the weakest of all lines, and now you may begin to understand why I refer to the closed-circuit lines as 'soft energy lines'. *The three strongest lines of a generator/alternator are identified in this book as the* **triple dipole lines**.

The total number of lines which exist within absolutely every magnetic field is a constant, but they do vary in strength and thus length. The magnetic flux density of the lines do vary, making it possible to detect each line, especially with iron filings; but eventually, the flux density of the <u>later</u> closed-circuit lines becomes undetectable by iron filings due to their weakness in the level of flux. The total number of lines and their positions have never actually been officially recorded, but it may now be time to explore this area of magnetics. This includes the total number of open- and closed-circuit lines. *The predetermined total number of magnetic lines of open- and closed-circuit lines which exist is set by nature and is what I refer to as the* **common magnetic total** *but is yet to be discovered.*

The exact number of 'other' open-circuit lines on each side of the three strongest lines is also a predetermined constant set by nature. Plus, the closed-circuit lines which radiate from pole to pole on each side of the major magnetic field become weaker in magnetic energy as they number or step away from the first strongest closed-circuit cage of lines. After the first cage of closed-circuit lines, the next cage is weaker, which will eventually be calculable in a percentage or common reducing ratio compared with the previous cage lines.

This link displays the total magnetic lines visible by iron filings: https://www.bbc.com/bitesize/guides/z3g8d2p/revision/2. And if the link is removed, simply search for 'iron filings sprinkled onto a magnet'. Also, if it cannot be found, can someone please create another link for the next edition of this book?

As previously described, the energy flow of a closed-circuit line begins from the first cage, just outside the major magnetic field or body of the magnet, but I name the area of all closed-circuit lines as 'soft energy lines'. The strongest area of soft energy of all closed-circuit lines is closest to the metallic body of the magnet with every other closed-circuit line of the soft energy, thereafter reducing in a ratio of flux density.

The natural length to which any open-circuit line travels then simply stops and exists in a natural or common length ratio to one another in length, depending on the total flux density of the magnet. Therefore, the predetermined length of two absolutely identical magnets (for their open-circuit lines) will be absolutely equal (that is, if it were possible to find two absolutely identical magnets).

I'll add here that each length (in a common length ratio) of the open-circuit lines that travel or flow and then simply stop will one day be possible to calculate by the common distance ratio formula, but for now, throughout the book, I refer to this predetermined length as the **common magnetic ratio distance**, which correlates directly with flux density and thus strength.

Each of the four lines in the original diagram are open-circuit lines travelling to a predetermined magnetic ratio in length away from the pole face and then simply stopping in accordance to the nature of magnetics, but the length and circumference of the four lines in the diagram exist in different pairs by size. Soon I will describe to you why each of **the four lines exist in pair sizes**, which also created the planets in pair sizes, clearly visible in the following planet chart link: https://goo.gl/images/zS1hij. This link shows the planets in pairs by size, but it only displays the eight largest planets due to the dwarf planets being so far away and orbiting in an extremely random order.

The original four lined diagram should display the other two open-circuit lines on each side of the centre line which generate the other two larger neap tides. These neap tides are therefore also in pairs and occur every quarter, but as stated, it is an extremely simple sketch.

I refer to the predetermined natural distance that an open-circuit line travels to throughout this book as the 'common magnetic ratio length of open-circuit energy lines'. I also refer to the predetermined distance of the gap between open- and closed-circuit lines as the 'common magnetic ratio distance' between energy lines. In closing, the total number of magnetic lines has never actually been explored in both open- or closed-circuit lines but will need to one day soon be uncovered.

The next few facts are the primary laws of magnetics, and the two most common laws are:

1. **the law of attraction,** where two opposite or opposing poles will attract each other, and
2. **the law of repulsion,** which states two like poles or the same poles will repel each other.

Our Earth has two polarities. North is above the equator, and south is below. However, the **north end of a compass needle is actually a magnetic south** and points north due to the law of opposites attracting, and the south end of the compass needle points north also due to the law of opposite attraction. Our Earth has two polar caps, and each hemisphere does possess a polarity, but our Earth also has a **twelve-inch band of neutral magnetic energy, which I refer to as the neutral magnetic zone**, (NMZ) in the centre of the equator that is actually more than just a neutral band of energy. As a compass moves across the equator, the compass needle will align in a contrary direction of east to west in the centre within the twelve-inch band (across the equator) where it should logically align in a north-south direction. *I refer to this twelve-inch band on our equator as the **twelve-inch neutral magnetic zone (TINMZ).***

Also, on the equator in the centre of the TINMZ, an egg will balance its point. In each hemisphere of our earth, water rotates in eddy currents or a vortex going down the sink, as well as in our oceans. These rotating gyres are generated and identified in science as the Coriolis effect.

Water gyres or eddy currents in the **northern hemisphere** spin in a **clockwise** direction going down the sink. In the **southern hemisphere**, they rotate **anticlockwise**. BUT in the **centre of the twelve-inch neutral magnetic zone** on the equator, water falls straight down the sink.

Also, just one day in our orbit, an egg will balance on its point but only for a few minutes. Tornadoes are not active on or near the equator within twenty-eight degrees. The egg phenomenon occurring every orbit is good evidence of the existence of a neutral magnetic zone within the energy of our solar system amongst the open or closed-circuit lines. This is due to our Earth orbiting through a neutral magnetic zone, but on which day I can't exactly remember because it hasn't been in the media for quite a while.

The solar system NMZ physically exists either in the middle of just one of the open-circuit lines drawn into the diagram or in between two diagram lines within that energy zone. It could be within the missing closed-circuit lines but does occur somewhere in the energy field of those lines which are generated by magnetic poles (to state the obvious). I would like to also add that it may be possible that the media have only balanced the egg in one particular hemisphere, and it may be possible that the egg will also balance on its point in the other hemisphere but six months away from where it is recorded now.

I would like to satisfy my own curiosity to know one day if a compass needle will align in a contrary direction simultaneously during the egg phenomenon as we orbit through our solar system

NMZ. However, I feel that our Earth's magnetic field may, in fact, be too strong to allow a compass needle to be affected on that day of the egg phenomenon by any solar system neutral magnetic zone, which may be present during our orbit.

The energy of closed-circuit magnetic lines on our Earth, which we identify and name as 'ley lines', flows from the south pole to the north pole throughout the crust and body of our Earth but also above the crust as it will within any magnet. The energy lines of any bar magnet (displayed in the iron filings link) will have closed-circuit energy lines projecting from the south end and the same closed-circuit lines entering in the north. The only place so far that a neutral magnetic zone has been located or identified in existence within the centre of any magnet is in the twelve-inch neutral magnetic zone on our Earth in the centre of the equator. But logically, *the neutral magnetic zone must obviously exist in the centre of absolutely every magnet and has gone undiscovered until now.*

With these simple facts in mind, I believe so far that we have only been shown the compass needle pointing east to west mostly within the TINMZ of our Earth, in the region of Africa, in social media. I do not remember ever seeing this compass phenomenon elsewhere on the globe other than Africa.

I theorise that the magnetic energy flow of any one closed-circuit magnetic line within our Earth moves according to the following descriptions:

a. The energy flow of magnetic flux particles **begins very gradually from the start of the South Pole in the centre of our Earth** and then flows through the six-inch portion of the southern NMZ and increases in both magnetic energy and velocity after leaving the zone.

b. The major magnetic field energy (in flux particle form) travels through the southern hemisphere of our Earth toward the South Pole and then flows as a number of closed-circuit concentric lines (magnetic ratio distance apart) away from the South Pole towards the opposing North Pole.

c. The energy (in flux particle form) within each line re-enters our Earth at the opposing North Pole and travels through the northern hemisphere (just as it will in a bar magnet) to the northern edge of the northern NMZ.

d. The magnetic energy then slows down again as it flows through that six-inch portion of the northern neutral magnetic zone, eventually slowing and flowing very gradually back to the centre of the north side of the centre of our Earth, completing the magnetic circuit of a closed line.

This, of course, also describes the magnetic flow of every magnet. Why should any magnet be any different to the field of our Earth?

The flow of energy across the twelve-inch zone is slower due to reactivity, which is because the energy of two opposing poles meet where it eventually peaks in the centre of the NMZ. The centre is where the closed-circuit energy lines begin, but the two polarities do NOT meet. (Why the reactivity of the compass needle occurs where the two opposing poles meet within a magnet instead of attracting is a bit of a mystery for me so far.)

The greatest observation we can make of the flow of magnetic energy or of closed-circuit magnetic lines is in the example of our Earth. It may be possible to learn more about magnetic energy by observing the energy associated with our Earth, especially on the equator or by measurements taken as our Earth orbits through each of the four lines within our solar system. By taking magnetic measurements on every day for twenty-four hours simultaneously in both hemispheres, we would be able to ascertain the definite existence of the four magnetic lines of the diagram and the distance that exists between lines, instead of the number of poles, which science records today, and I'll tell you more about that at the end of my book. (We could, of course, also start by measuring the magnetic ratio distance between closed-circuit ley lines on our Earth or uncovering more about what is occurring in the TINMZ on the equator all around our Earth).

The compass needle aligning perpendicularly, the egg balancing on its point inside the NMZ, water dropping straight down the sink, and the absence of tornado activity on the equator only occur in the TINMZ on our Earth. These are all strong indicators of the presence of a 'neutral magnetic energy zone', which will also obviously exist within every magnet (in a narrow strip, especially seen by our Earth's energy field), which has so far gone undetected by science.

The NMZ of every magnet would also exist at a specific distance ratio (according to the magnetic flux density of each magnet), which would highly likely be calculable by the magnetic ratio distance, which will also someday have a formula to make it possible to calculate each zone's distance or size in proportion to the remainder of the magnet in a ratio to distance and strength. *This formula will at best be identified today as the* **common magnetic neutral energy zone ratio.** *Through my observations of our Earth's twelve-inch neutral magnetic zone, I identify the neutral magnetic zone of every magnet as the* **common neutral magnetic zone.**

Each of the four lines of the diagram are all open-circuit magnetic lines, with each line possessing a polarity in a clockwise direction, north and then south, south and then north, and I'll explain the peculiar or unexpected irregularity of these polarities later, which is very important. The extremely simple diagram sketched as four lines and a reactive centre sphere does not display any of the 'other' open-circuit lines or any of the closed-circuit lines that will also exist as they radiate out into our solar system as magnetic energy lines.

It's very strong evidence when the egg will balance on its point during each orbit (for just a few minutes) that a neutral magnetic zone also exists within our solar system on one day of our orbit, but it will also help us know if the compass needle will align perpendicularly and simultaneously during the day of the egg phenomenon or if there are, in fact, two days when the egg phenomenon will be possible, one day in the southern hemisphere and one day in the northern hemisphere. I say this because there are two days when opposite polarities change in our solar system. Remember that the lines have two different sizes. One of those days when the polarities change may, in fact, be too weak to be able to support the egg. It may also be possible that there are four potential days for the egg phenomenon to occur. Because the egg is supported as it stands up on its point, the neutral magnetic zone may be some energy other than neutral. The same must be said for a neutron of an atom, and I'll also get to that later in the book.

These four lines of the diagram exist in two different sizes, and later, I'll also show you the solid evidence which supports this. *As mentioned above, if the lines are in two different sizes, then we may have overlooked the fact that there are other weaker egg phenomenon days.* The four lines of our sun, which radiate out across our solar system in the diagram, are north and then south and south and then north in polarity; and during our orbit, our Earth rolls through an area where a south-north polarity meets. And just like our Earth has an NMZ (where a south-north polarity meets), so too does our solar system. The meeting of a south-north polarity (as it occurs on our Earth) may be very likely why the egg phenomenon occurs on our Earth. It may also be the fact that the closed-circuit lines meet in the centre.

As we already know, our compass points to magnetic north, but our Earth also possesses a true north. Magnetic north and true north were correct in 2015 here in Australia, according to the map surveys, but magnetic north is said to move westerly by 0.04° every five years. Regardless of how fast it moves, anyone who does bushwalking or orienteering or sailing can tell you the important fact that magnetic north is always on the move.

I would like to make a safe bet and say that, during the last 500 years of our Earth's orbit, magnetic north has moved from about eight degrees left of true north, through true north, to about eight degrees right of true north, although I must inform you that science only discovered the theory of a meandering magnetic north in 1831 and the actual compass was invented in around 200 BC. I would like to make another safe bet that the movement of magnetic north is caused by the stronger magnetic field of the neighbouring solar system(s) or maybe when their planets align in our direction.

Knowing that magnetic north is constantly on the move and also knowing that magnetic lines are fixed in place with no wavering or natural distortion, it is also very logical to say that the ley lines of our Earth or meridian lines, which are the closed-circuit lines of our Earth (as a global magnet), are also on the move but radiate from a fixed place from the pole. So when somebody says to you, 'This exact location in our area is a ley line or a meridian line which is very magnetic,' you can correct them and tell them why it doesn't stay in that one place if it is a ley line or meridian. It may just be that the area is simply magnetic, and I'll tell you more on that soon.

'Meridian line' is the slang term used to describe an active ley line location, but I would like to point out that they are constantly on the move due to magnetic north and south always being on the move, a point of science which has been grossly overlooked. This would, of course, also make it just about impossible for science to measure the distance between the ley lines of our Earth (unless, of course, they could measure the precise speed of movement and the amount of time it takes to move that distance).

In later chapters, I'll further define the term 'neutral magnetic energy', which quite possibly could be very active rather than neutral, especially in an atom. Science has not really explored the NMZ on the equator to any great extent, and it could be that the magnetic energy on the equator is very active in an upward direction, which will not be visible on the compass. This may be why the egg stands up?

In a later chapter, I make a comparison between the energy of the NMZ and the energy of the neutrons of an atom, which are recorded in a neutral polarity. And again, maybe the neutrons could not possibly have a neutral energy; after all, they are bonded and orbiting with protons (of a positive charge) within the nucleus of the atom. *Taking NMZ into consideration, we can clearly observe the presence of three energies within any magnet; and from the elements of an atom, we can also observe three energies of any atom.* The energy within each atom's nucleus is interacting with the neutrons, which make it possible for them to be able to stay in orbit amongst the protons and electrons. If the nucleus of an atom was the size of a basketball, the electrons would be thirty-two kilometres away. Later in this book I will discuss the energy within that region of empty space.

I honestly feel that the compass pointing east to west within the centre of the equator (in a contrary direction), the egg balancing on its point during orbit, the water going straight down the sink, and the inactivity of tornadoes on the equator all within the neutral magnetic zone are only an enigma so far for science today, which may, in fact, lead us all to an answer that will help us better understand the energy of a neutron or an atom's nucleus.

Recently, I read a description for magnetic energy as a 'force which is 137 times stronger than gravity'. Closed-circuit lines surround every magnet and reduce in their level of magnetic energy as they number or step away from the major magnetic field. The two strongest closed-circuit lines are positioned on each side and closest to the major magnetic field or metallic body of each magnet. The distance within each magnet held between each of the open- and closed-circuit lines is equal, predetermined, and dependent on flux density of the magnet, set by nature at what I refer to as the common magnetic ratio distance.

Each of the open- and closed-circuit lines (at the predetermined MRD) could be numbered in flux density from the highest to the lowest as they number or step away from one another, with the centre open-circuit line as number one. Every other open- or closed-circuit line after the centre line could also be numbered in sets because they exist in mirrored pairs, beginning with the pair of lines numbered 'B-L1' and 'B-R1' (left and right). This could eventually be correlated to our Earth's ley lines. I cannot number the total of lines of a magnet today in this book because the total number of magnetic lines of every magnet have gone unrecorded by science thus far, but it may now be time to pursue for some more exploration. How I know this ratio fact above will be revealed later.

Our Earth has two magnetic polar caps with open-circuit lines projecting from the North and South Poles, travelling and flowing from each pole to a point and then simply stopping at a predetermined distance. When comparing our Earth to a simple bar magnet, the inner core of our Earth is recorded as the source of the major magnetic field (as the body of the magnet) and the ley lines of our earth are the closed-circuit lines of the magnet. Closed-circuit lines, identified as ley lines, in our Earth's crust or meridians flow from pole to pole. The centre core of our Earth has a magnetic field but so too do the other layers, in the middle core and outer crust, all possessing closed circuit lines.

The closed-circuit lines as ley lines of our Earth flow through the crust and above our Earth, with the energy field of our Earth extending beyond our Earth's boundaries into space, beyond the stratosphere. The outer energy field of our Earth is represented by the blue hue which surrounds our Earth, seen in photos returned by NASA from space. I will tell you more about that energy of the blue hue later in the book. The weaker closed-circuit lines of our Earth as ley lines are beginning to look a great deal to me like 'gravity', which is said to be 137 times weaker than magnetic energy (here on Earth), but I'll also tell you about my theory on gravity involving the closed circuit lines later in this book.

The major magnetic field or the body of the magnetic field in regard to our Earth is recorded by science as being emitted from the centre core of our Earth, which consists of molten lava, according to the greatest source of science, NASA, who states that there are a recorded total of three cores within our Earth. The crust is the outer core, the inner core is referred to as the mantle or middle core, and the centre core is the molten lava core, which is recorded as the major source of our Earth's magnetic field and emits a frequency of just 8–12 Hz. This is, of course, strong evidence that magnetic energy has a pulse or exists in a somewhat fixed frequency, but more on that subject later as well.

The molten centre core (just like any bar magnet) possesses every one of our Earth's closed-circuit ley lines as the major magnetic field. The closed-circuit lines project from the South Pole and flow to the North Pole, and they reduce in energy and flux density as they number or step away in cages from the major magnetic field of our Earth. Remember that the major magnetic field or the magnetic body of every magnet contains the next strongest magnetic energy whilst every open-circuit, plus every closed-circuit, line flows through it.

The magnetic field of our Earth, which extends out into space, is described as the shield surrounding our Earth which is a barrier to our sun's solar winds and UV rays. The first cage of closed-circuit lines of our Earth also have slightly weaker lines again on the outer side with gaps between them, which can be measured in the common magnetic ratio. There are strong indications that these next set of lines or the next cage consists of secondary weaker ley lines outside or each-side of the primary lines or primary cage. A third weaker set of lines also exists outside the secondary lines or cage, and a further set of lines also exist outside of those again. The total number of sets of lines or cages is seven. Our Earth's magnetic field would be the best place to study these lines due to the sheer size of our Earth and it being the greatest example of magnetic energy, making the measurement of the MDR possible, which is the distance or gap between lines.

Logical evidence of a greater number of lines than just ley lines has been observed over recent years while observing homing pigeons (as well as other animals). Homing pigeons and other animals are recorded as **having the ability to navigate their way** by using our Earth's magnetic energy lines. I theorise that homing pigeons which are **seen walking back and forth in the 'home roost'** do this whilst physically feeling the presence and flux densities of the magnetic lines present there. They then commit the reading (taken of each magnetic energy line) to a physical body sensing, more so than just a memory recall. I am trying my best to say that the pigeon physically senses the magnetic line, and when they do, it activates a part of their body more when the energy is at its strongest. Later in the book, I move into a theory that magnetic lines possess a frequency, and this will make it very possible for homing pigeons to feel lines and predetermine their position on the Earth plane. Each home roost line to the homing pigeon will be very much like a familiar smell will be to a dog, who finds his way home by following the stream of the familiar smells with his nose.

Recently, I spoke to my father about homing pigeons that fly over one particular point on his house. He lives on a rural property and said that recently, when they fly over his house, whilst he was sitting on his porch, he often saw a single homing pigeon (notably having the same varying colours) flying over just one point of the house. Each time one flew over the identical point, it grew more and more obvious that these birds followed one common path. (I suggested to him that it was magnetic ley lines.)

After speaking to him, I then saw a TV program where the flight path of homing pigeons was recorded on a computer screen using a small global positioning system (GPS) device attached to their backs. When let loose, the pigeons began their flight by circling at first for a minute (as they normally do) and then following what I can only assume were the familiar, homely smells or frequencies of the ley lines and other finer lines which are close to the frequencies of the home roost's magnetic ley lines. The lowest frequencies are only possible to be sensed by a homing pigeon with the flight paths recorded by the GPS mapping on the screen data.

I assume that the pigeon's circle for a minute or so whilst physically sensing the ley lines or maybe also other weaker lines on either side of the ley line. I have recently concluded that there may in fact be three lines at every ley line, for the logical reason that ties in with the three lines from college. They may also be sensing the width of each ley line, which might help measure the frequency (sensing familiar smells, so to speak), which is what the pigeon does daily/hourly when roosting, whilst on foot, moving from side to side in the home roost. I further theorise that the pigeon is always working (whilst also flying), memorising indicator widths and frequencies or familiarising itself with homely smells whilst generally flying around, especially when circling the home roost every afternoon. Remember the lines are always on the move and the pigeon may in fact be aware of this.

It is a well-known fact that the person who owns homing pigeons must let them out every afternoon to simply fly around the home for exercise, but I feel it is also for sensing lines that may be on the move. Whilst flying around the home, the pigeon is physically feeling the ley lines around his home whilst higher up in the air than the roost itself. The bird paces up and down in the roost because they are obsessed with working and staying in tune with the local ley lines, always sensing magnetic lines with their body or maybe their brain. At some stage, the bird will have memorised the ley lines which flow in the residence area to where it is released, maybe whilst on other flights, or it may be that the homing pigeon is always flying back and forth across the paths of lines, memorising indicator widths or frequencies.

I myself had, on occasion, felt the magnetic energy of one of the diagram lines. It made me suddenly become off-balance on my feet for just a moment, around the time of a king tide. As I also previously mentioned, it can often make the elderly feel off-balance.

When finally familiarising itself with the local lines by sensing them during that minute of circling, the pigeon may then choose to follow a recognisable ley line that it knows is the return path to the home roost or will lead to the home roost's ley lines, recognisable to the bird's brain by a physical memory. Please remember that the ley lines of our Earth are always on the move, and this may be the reason the bird is always sensing and recording ley lines.

Of course, when the pigeon recognises the home roost energy line, it will only be a matter of physically following the familiar energy of the line with its physical body or brain by just simply tracking the line as it flies. The magnetic lines are obviously a physical sensation to the pigeon, just like how the four lines of the diagram can cause an unsteadiness to the elderly on those days when the diagram lines pass our Earth whilst we orbit through one. Because a homing pigeon can follow a greater number of choices as the 'path home', it makes me wonder whether there are a greater number of lines.

I am trying my best to gently say that maybe the ley lines of our Earth (which will be the closed-circuit lines covered in iron filings) have narrower and/or weaker lines on either side of them. At this stage, who could honestly know besides the animals that use the lines to navigate? Of course, there will be many indicator strengths as choices the homing pigeon will recognise to enable them to return to home, beginning with being able to feel and follow even the narrowest or weakest and lowest strengths.

When considering a whale that navigates its path around the world by the use of ley lines, you can only imagine that this may be one of the reasons why a whale calls to its young, which may be in training for recognising and following the ley lines. Recently, there was a new science program that came to us on pay TV. This program spoke about a little brown bird that looked a great deal like a sparrow to me. They had been tracking the migratory path of the bird and discovered that it flew 4,000 miles south during winter, but the bird wouldn't leave during times when the hurricanes were occurring in that southern area, before the storm occurred 4000 miles away. The scientist on the program said that they could now predict when a bad hurricane season approached the southern area because the sparrow would not migrate from the north 4000 miles away before the hurricanes occurred. The scientist could not figure out how the sparrow knew that there would be bad hurricanes before they occurred. I suggest that the bird could feel magnetic lines in a stronger field strength before a bad season. I will also enlighten my reader later in the book about Tornados and Hurricanes.

Recently, an indigenous Australian woman told me that her tribe's area of land was notoriously known for its magnetism and meridian lines. She asked me for its cause after hearing my theory on moving meridian lines. I suggested to her that their area of land was very obviously the ancient location of an enormous magnetic meteorite landing with an extremely high content of magnetic rock, coming to rest during a period when our Earth was very soft. The meteorite had since melded into the landscape and also gathered other magnetic rocks and dust around it.

Again, in regard to open-circuit magnetic lines, apprentice electricians are taught in **magnetic theory** that every magnetic pole has three open-circuit lines projecting from absolutely every pole face which are the strongest in flux density of any lines. Just those three strongest lines alone generate about 75% of the total during power generation, but in college, those lines were simply referred to as 'three open-circuit lines' that I call 'triple dipole lines'. We were taught that the centre line is of the highest flux density, with the next outer line on each side being just slightly weaker in energy. The remaining approximate 25% consists of **other open-circuit lines**, plus the **total of closed-circuit lines**.

The other visible open-circuit lines (according to the earlier iron filings website link) are weaker in comparison to the three strongest lines; plus, every other closed-circuit line is weaker again, until they cannot hold any iron filings. The image in the 'iron filings website link' only displays the stronger lines which possess the highest level of flux density, but as stated, there are a great number of lines which also exist within every magnetic field that will NOT hold iron filings.

The closed-circuit lines of our Earth as the ley lines would be the greatest way to observe closed-circuit magnetic energy lines. The magnetic energy of a magnet is recorded at 137 times stronger than gravity, but the closed-circuit lines of our Earth are very likely as low in energy as 137 times weaker than the open-circuit magnetic lines of any magnet. What I am trying to say by this is that the closed-circuit lines, as ley lines, make up our Earth's general magnetic field in the crust but may, in fact, be the energy which produces our gravity field due to the great number of ley lines, plus other lines which may exist on either side being roughly 137 times weaker than the open-circuit lines. (I'll show you more evidence surrounding this subject later. Also in that later chapter, I'll describe to you precisely how every magnetic line of our Earth was actually first initially created.)

In closing, I would like to remind you of another childhood experiment that we all saw. By stroking a piece of mild steel with a bar magnet, it will generate or create a new magnet. The polarity end of the stroking magnet being used (either north or south) will predetermine which polarity is generated or created at each end of the mild steel. Also, if a weaker magnet is stroked by a stronger magnet in the same direction of the magnetic particles, it will induce a stronger magnetic field into the weaker magnet. <u>When using this stroking method</u>, you must only stroke in <u>one direction</u> to enable the proper transfer or inducement of magnetic flux energy particles.

An electromagnet can also be created by wrapping an insulated wire around a non-ferrous iron bar (which will not retain magnetism) and by charging the wire's terminals with DC electricity. The magnetic field created by the electrical current induces magnetism into the iron bar that will not retain the magnetism. This produces a man-made electromagnet which will NOT retain any magnetism after disconnection of the electrical source. The induced magnetic field will collapse the moment disconnection occurs.

The direction that the coil winding is wound will predetermine the temporarily induced north or south pole end of the new magnet in the iron bar. Reversing the electrical polarity at the terminals will also reverse the temporary polarity, as well as reversing the direction which the coil winding has been wound. Remember that the magnetic energy will NOT be retained once the supply is disconnected, and this is why non-ferrous iron is chosen for an electromagnet, making it possible for a scrap metal yard to drop the steel after it was picked up by simply switching off the electrical supply.

An electrical transformer is also made up of non-ferrous iron laminations and functions under very much the same principal above, when the electrical field is produced or induced, producing a magnetic field into the transformer's secondary coil winding. The transformer uses iron laminated steel sheets as the core so as to reduce the generation of eddy currents. (I will have more on eddy currents soon.) A transformer can only function in an alternating current (AC) electrical circuit when the alternating current moves through the zero-volt axis over and over again, causing the collapse of the magnetic field over and again. A transformer in a DC circuit has a constant supply and cannot produce a collapsing field unless you switch off the supply, which is the primary method of function of a car coil.

The following describes the transformer's mechanism but is not essential to understand:

a. When the positive cycle of the alternating current enters the primary coil, the electrical current flow charges the primary coil winding with a magnetic field in the first half of the AC cycle, which is a positive electrical charge, generating a magnetic field or polarity from the direction of electrical flow within the winding.

b. The AC current flow then reduces as it flows through the zero-axis point (zero volt), which will collapse the magnetic field of the primary winding, and the collapse of the magnetic field will generate electrical inducement into the secondary winding via magnetic inducement energy. Remember two energies functioning together as one.

c. The AC electrical flow then changes to the negative half of the AC cycle, and the magnetic field is induced again into the secondary coil winding, and the magnetic collapse cycle starts generating electrical inducement into the windings again.

d. It's the collapse, of the magnetic field which produces movement of magnetic particles, thus inducing or transferring a magnetic field into the secondary coil winding of the transformer. The collapse of the secondary winding creates movement of magnetic particles which, in turn, generates the output of the transformer.

An electromagnet will generate a magnetic field using a constant supply of DC electricity, and a transformer functions under a very similar principle of inducement. It's important to understand that a non-ferrous iron core which does not retain magnetism is utilised in each electrical application.

The ignition coil of a car functions under the same principle as a transformer, and in fact, it is a transformer, although it is DC electricity, which is supplied to the car coil and switched on and off. The switching (on and off) by the points of the car produces an alternating field. A constant supply of DC is applied to the car coil, but when the distributor points open (in an older car), it collapses the magnetic field of the car coil winding which, in turn, produces movement of a magnetic field and induces a magnetic field into the secondary winding of the car coil. The secondary winding produces the high-voltage output to the spark plugs. The secondary winding simply has a great number of turns in its wire coil.

The collapse of the magnetic field in the primary winding generates movement of magnetic flux particles and a spike of high voltage in the secondary winding of the car coil. Most importantly, it's the opening of the points (in older cars) which collapses the magnetic field, thus generating a movement of particles in the secondary coil, and the opening and closing of the points creates an alternating pulse. The fact that the secondary coil has a greater number of turns in a smaller gauge wire is the reason why a higher secondary voltage is produced. This principle of charge/discharge also applies to the ballast of a fluorescent tube light when the starter strikes.

The points of an older car are, of course, the ignition switch mechanism of the car whilst most modern switching of distributors today is carried out electronically using a circuit mechanism called the Hall effect, which is actually important and discussed in a later chapter.

It's not important here to know every part or mechanism of a transformer, but it is important to understand that it's the law of magnetic inducement applying to this mechanism which produces the energy and also that a non-ferrous iron core is also used because it won't hold a magnetic charge, as in electromagnets. The magnetic field is NOT retained during inducement into the iron in either of these electrical applications, thus not retaining a magnetic memory which is why ferrous-iron is chosen for this application(s). Also, when a piece of mild steel comes in proximity to a magnet, it will be submitted to a mild charge by magnetic proximity inducement, but it will NOT retain the magnetic energy after the magnet is removed. This functions in a similar way to a transformer.

The inducement occurs due to magnetic 'proximity' inducement, and the other metal should possess an ability to retain magnetic energy, and mild steel is just one of those. Induced proximity magnetism is important so as to understand where our moon's gravity is induced with magnetic energy by each diagram line as it orbits through each line every three months, generating each of the three strongest tides, but it does not retain that level of magnetism. The moon does NOT retain the magnetic energy induced by each of the diagram lines when each of the three strongest tides is generated. Inducement by other planets generate extreme weather events, but the collapse of a magnetic field which generates high-voltage spikes does not apply to our other planets and is not relevant.

Shall we take a look at the details of the three strongest tides? **The following describes how the four diagram lines generate king tides, plus how each of the two other outer lines (not included in the simple sketch) generate each neap tide.**

According to science, daily tides (generated every six hours and fifteen minutes), plus each king tide (every three months), including each of the two neap tides (occurring on each side of a king tide), are all generated by our sun and moon's gravitational energy (inducement of the waters by the gravitational energy; remember, everything is magnetic). Science records each king tide as the strongest of the three, generated in the middle of the three strongest tides every quarter of a year. *(I use the term 'quarter of a year' for the three strongest tides so as not to confuse it with the term 'every three months', which is used in another theory description later.)*

The diagram is a very simple sketch which does NOT include the flux density and the other two open-circuit lines or any of the closed-circuit lines, but from the information learnt as an apprentice electrician, we all now know that the four diagram lines possess a high percentage of the total energy and also that magnetic energy is said to be 137 times stronger than gravitational energy. To be totally accurate, the diagram should also include all the other lines on either side of the centre line, especially the other two slightly weaker outer lines.

Each of the two neap tides occur as a single tide before and after each king tide, and each neap tide is just slightly weaker in amplitude compared with each king which occurs at that time. The ratio of king tide amplitudes compared with neap tide amplitudes exists, is the same as in the lesson from college where the centre line amplitude is highest compared with the other two slightly weaker outer line amplitudes. The three largest tides and the magnetic energy ratio of the lines are a direct reflection of each other.

The following briefly reiterate the outline of just one of my theories, which is the cause of the <u>generation of each king tide</u> occurring as a result of our moon being temporarily induced with magnetic energy from each passing line of the diagram when our moon orbits through each line. The magnetic energy of each diagram line induces a stronger magnetic energy into the moon, thus amplifying the moon's normal gravitation energy, and the stronger magnetic energy of each line is NOT retained by the moon. The diagram is missing each slightly weaker line on each side of the centre line, and each neap tide is an amplified energy by each missing line when our moon orbits through each missing line, and the amplification occurs in exactly the same manner as each king tide.

My theory describing the generation of each king and neap tide revolves greatly around a few previously mentioned facts:

- Every magnetic pole has the three strongest lines, with the strongest in the centre.
- Each of the other two outer lines positioned on either side of the centre line is missing from the diagram and slightly weaker in magnetic energy than the centre line.
- Each of the diagram's four lines is primarily magnetic energy with a suggested force that is 137 times stronger than gravity. The diagram lines are weaker than 137 times due to their distance from the magnetic poles of our sun.
- The diagram's four lines make one complete rotation every four years with a clockwise polarity of north and then south, south and then north.
- Each of the four lines spans about fifteen billion kilometres across our solar system, implying that they will be weak in magnetic energy compared with the magnetic energy of an average magnet of normal open-circuit magnetic energy here on our Earth.
- Each king tide (generated every quarter) is a side effect of amplification by each of the four lines when our moon orbits through each diagram line; plus, each neap tide is generated under the same principle.
- As an extremely simple sketch, the diagram does not display the 'total number of lines', including the other open-circuit lines, plus every closed-circuit line which will exist within our solar system.
- Each line's magnetic energy which is slightly stronger than our moon's gravitational energy generates magnetic proximity inducement, resulting in the moon generating each king tide, plus each neap tide.

I identify the three strongest tide lines in my theory of 'amplification of our moon's daily gravitational tide energy' as the source which generates each king and each pair of neap tides, and I call them **Cox-factor tide lines**. *I also identify each of the three tide lines occurring in our solar system singularly as the <u>king tide centre line,</u> which generates king tides, and each* *of the other two outer and slightly weaker lines generating each neap tide as the <u>neap tide outer lines</u>, which are missing. This new identification is important in a later chapter on tornadoes and cyclones.*

In my magnetic theory class in trade college, each of the three strongest lines (as the Cox-factor tide lines in our solar system) were simply referred to as 'open-circuit lines'. I now identify those three power generator lines as the <u>triple dipole lines</u> and singularly with the centre dipole line as the <u>centre triple dipole line</u> and each outer line as an <u>outer triple dipole line</u>.

The simple sketch does not express the level of flux density for each centre king tide line (centre Cox-factor line), and in a moment, I'll describe exactly why the circumference of each line and the strength and especially the length of each centre king tide line of the diagram are important.

The easiest way to possibly read the tides is by a 'tide book', where you can highlight every month's highest tide to then be able to see the four highest yearly king tides, as well as each slightly weaker pair of neap tides before and after that king tide. *It is of utmost importance to note that a king tide occurs exactly during each change of the season. From each quarterly group of markings, you will notice with some attention that the tide book will now clearly show that each of the four king tides are also generated in separate but distinctive pair sizes. The king tides occur in two separate amplitudes or heights and strengths, which I refer to here (for ease of reference) as the* **pairs of king tides**. *These pairs of king tide lines (Cox factor) exist in two distinctive amplitudes, which generate two distinctive king tide amplitudes due to the Cox-factor lines existing in pairs by amplitude or energy.*

There are a pair of 'large-circumference' Cox/king tide lines and a pair of 'smaller-circumference' Cox/king tide lines, with each pair of lines existing in different strengths, lengths, and circumferences reflected in the two different tide sizes. Soon I will show you how these lines created each of the planets, which also exist in pairs today, but the simple diagram sketch does not show the difference or pairs of lines. It will, however, be possible to understand the pairs of lines when I also show you how our sun was first generated and formed.

The pairs of Cox/king tide lines generate 'pairs of king tides', which occur every six months because the 'pairs of lines' are opposite to one another in the diagram in positions that you would notice if the sizes were drawn into the diagram. The fact that the pairs of king tides are on opposite sides of the diagram is the reason why our Earth undergoes two spring tides but singularly in each hemisphere.

Soon I will also give you finer details of another of my major theories which describes precisely how each planet was formed on each line, but for now, I'll tell you that the 'pairs of size lines' which generate pairs of king tides formed 'planets paired by size', which are clearly visible in this planetary chart link: https://goo.gl/images/46VoDG. Again, if the link disappears, simply search for 'ascending planetary chart'. I'll describe how all this occurred in what I refer to as 'slow reveals' in simple details soon.

To help you understand the next theory, let's now take a quick look at a few simple facts about electromagnetic energy (EM energy). EM energy is actually two energies (electrical and magnetic) which function together as one. It may be easier to understand this relationship if you were to visualise them on a set of imaginary scales with magnetic energy on one side and electrical on the other. In the most natural state, EM energy exists as a very high percentage of magnetic energy, which is the case in the four diagram lines. When electricity is generated, the natural state as EM energy is altered, the scales tip, and EM energy will now exist in a very high percentage of electrical energy.

As magnetic energy or as electrical energy, these two energies (on either side of the scales) could each technically still be identified as electromagnetic energy. These two energies can never be separated or parted from each other.

Because of the functioning of **electrical energy together with magnetic energy**, the poles of every magnet also possess an electrical polarity. Magnetic north is a negative electrical polarity; magnetic south is positive electrical polarity. In a direct current (DC) electrical circuit, one battery terminal is a positive polarity, and the other is a negative. DC electrical current flows from the positive terminal directly to the negative terminal without fluctuation which occurs in the form of alternating current electricity. Direct current is best described on a graph where the level of electricity is displayed as a straight line, running from left to right above the zero axis. If it were twelve volts (car battery voltage), then the line would be straight across to the right at a level on the graph above zero volt where there is twelve volts. Negative twelve volts DC, the line will be displayed from left to right below the zero axis in the negative scale.

As stated above, north of any magnet is a negative electrical polarity, and south is a positive electrical polarity, with electricity or electron flow occurring from the positive to the negative terminal. However, magnetic flux energy particle flow occurs from the magnetic south pole to the north pole, in the reverse direction to electrical flow, which is electrical positive to electrical negative. Electrical energy particle flow (electron flow) and magnetic energy particle flow (flux particle flow) occur in two opposite directions and are only possible due to the laws of electromagnetic energy functioning together as one but as two different energies.

Recently, I saw a documentary on science where the quantum physicist could not explain why particle flow within subatomic particles existed in two different directions. I suggest now that the law above might be a good place to start any investigation. (This is merely a suggestion. I am not implying that I have any greater knowledge.)

Electromagnetic energy as a very high percentage of magnetic energy is, of course, identified as a magnet where the scales are tipped to the side of magnetic energy but at the same time also possessing electrical energy; and despite being identified as a magnet, it is technically still just EM energy being emitted by the magnet. From this description, you can clearly see why I refer to electrical energy as an 'altered state' which is produced after electrons begin to flow in the coil of wire as it passes the magnetic pole, which has magnetic energy flowing from it. At this exact point, as the coil passes the pole, the coil of wire is deemed as 'cutting flux' whilst the electrons begin to jump which **will not jump in insulating materials deemed as insulators.**

The electrons can only jump because gaps exist in the next atoms' outer orbital, and this is due to a uneven number (creating gaps) of electrons existing in the outer orbitals of conductors. The electrons cannot jump in the example of insulators due to the absence of gaps in the outer orbitals of insulators. There are an even number of electrons in the outer orbitals of insulators.

The only place I can think of when or where these two energies may be in balance is in an electromagnet where a wire is wrapped around a non-ferrous iron core, connected to an electrical source, producing a magnetic field, and the magnetic energy is temporarily induced into the iron core generating a magnet from the non-ferrous iron by electrical means. *The magnetic energy is being induced into the iron core, generated by the voltage within the coil of wire; and when the induced magnetic energy occurs between two objects which are in proximity, I refer to this induced mechanism as **induced proximity magnetism**.* The primary use as an electromagnet is in a scrap metal yard where the supply voltage can be switched on or off to pick up and drop off the metal.

When placing a compass alongside a cable with a direct current flowing in it, the compass will align in parallel with the length of the cable in a south-to-north orientation caused by the presence of magnetic energy flux particle flow which is inducing the magnetised compass needle. This is a good example of a stronger magnetic field (the cable) inducing its proximity magnetism into a weaker magnetic field (the compass).

The following is a brief description of power generation including the direction of generated current flow, which will be important in a later part of the book describing the direction of gyres. The output of direct current electricity, which is the output of a generator, is directly proportionate to the flux density of the dipoles of the generator with the three strongest dipole lines generating 75%. Output by either machine is by a generator (DC) or an alternator (AC) and is measured in volts and amperes, which is identified as the 'rated wattage output' of the machine.

Alternating current (the output of an alternator) is also measured in 'frequency' and may differ from the revolutions per minute (RPM) of the machine. **Absolutely every AC energy possesses a frequency.**

Direct Current for Generators and Alternating Current for Alternators

A machine which generates AC is called an alternator, and a DC machine is identified as a generator. The following facts about Fleming's law help us understand why the direction of current flow can also differ during the generation of electricity. *This is important in my later theories, but if this chapter is too difficult, then take a break; and when you come back, read it one line at a time as I do, but please enjoy the book as much as possible.*

Taking a coil of wire (with the circle of the coil facing you) and then placing the **north polarity end** of a magnet **into the coil,** the movement of magnetic energy particles through the coil forces movement of electrons (which is altering the state of the EM energy), and the generated DC electrical flow in the wire coil will only flow in one direction of the coil. Reverse the magnet and then place the **south polarity end into the same side of the wire coil**, and the direct current electrical flow will also be reversed to the opposite direction of the wire coil. If you then replace the **north polarity into the back side** of the wire coil, the direct current electrical energy flow in the wire coil will move in the opposite direction to the **magnetic north entering from the front**.

These facts help us understand that *different magnetic polarities entering from different sides of the coil will generate current flow in a different direction for each change* due to magnetic particle energy flow occurring in a different direction for each of the two magnetic poles, forcing electrons to then flow in two different directions in the wire coil. When alternating current is generated in an alternator, it uses the laws of wire coils and magnets above, but the only difference is that the wire coil is the part of the alternator which is moving, whilst the dipole magnets are stationary.

Alternating current displayed on a graph will travel in a sine wave where the line voltage will start from the zero-volt axis and then curve upward in the shape of a hill when flowing back down to the axis. The line will cross the zero axis and then follow the shape of another hill, upside down, in the negative voltage zone; the line is always moving forward to the right of the graph because left to right is the period. When the line on an AC graph starts from zero on the axis and begins to travel upward in the form of a hill, it is increasing in the amount of volts; at that moment, within the alternator machine, the wire coil is moving through the magnetic field of each pole. When the wire coil is passing the middle of the magnetic pole, the peak of the hill or the peak voltage is reached, and the coil will be passing the three strongest dipole lines. After the coil of wire passes the pole and magnetic field lines, the volts on the graph will descend down the hill, so to speak.

In a two-pole alternator machine (when starting from the position halfway between the poles), this is where the zero-volt output is on the graph. Just another quarter turn or ninety-degree mechanical turn of the machine will be where the wire coil is passing the magnetic poles at the peak voltage. At a ninety-degree turn, the wire coil will have rotated to a point which generates peak output, which will be just half a cycle (or half a hill) on the graph, having travelled from zero to peak voltage (or half a hill).

As the coil has passed the centre of the pole face, the peak volts will begin to decline with the other half of the hill on the graph, now appearing as it moves down (the hill) towards zero volt again. When the line of the graph reaches zero again, the coil is now positioned on the opposite side of the machine, to where it began between the two magnetic poles. As the coil then moves towards the opposite pole face, the output volts begin to move towards the negative peak in that half of the cycle (towards the peak of the hill but in the upside-down position). This is due to the law of the magnet and coil being reversed in the original description of the laws above (that is, the opposite pole of the magnet re-entering the wire coil).

In a two-pole alternator, when the wire coil rotates one complete mechanical turn, this is referred to as 'one complete mechanical revolution' or, more importantly, 360 electrical degrees as one electrical cycle. This is when the line on the graph has completed one sine wave or two complete hills. One hill will be above the zero axis, and the other hill will be below the zero axis (upside down).

As the wire coil continues to spin and completes a number of mechanical revolutions within a two-pole alternator, the line on the graph will continually draw sine waves (hills) up and down in both positive and negative sides of the zero axis. Each hill (either upside down or right side up) is referred to as 'half an electrical cycle' and is measured in cycles per second as the frequency. Here in Australia, the frequency of the electricity is fifty cycles per second or fifty

Hertz. When the motor which is driving the alternator is spinning, it is measured in revolutions per minute. RPM and the frequency in cycles per second are two different means of measurement.

When an alternator has four poles, it will have two sets of dipoles within one machine; and just one complete mechanical turn of the machine, which is 360 mechanical degrees or one complete revolution, will produce 720 electrical degrees or two complete 360-electrical-degree cycles or two complete sine waves. This is because the coil of wire has passed four magnetic poles. In a four-pole machine, the magnetic poles will be positioned in north and south, north and south polarities (which differs from the poles of our sun). Just 180 mechanical degrees or half a mechanical turn of a four-pole machine will produce one complete electrical cycle or 360 electrical degrees because the wire coil will have passed a north and a south pole by the time it has completed just half a mechanical turn. In one complete 360 mechanical turn of a four-pole alternator, the electrical cycle will have been completed twice or have completed 720 electrical degrees or two 360-degree electrical cycles or two complete sine waves.

Compared with the laws of coils, the magnet will theoretically have been moved into the wire coil twice and reversed twice, causing the electricity to flow within the coil, reversing in flow in each direction twice. The only difference is that the wire coil passes the magnetic pole on the outside of the coil in the example of the alternator, instead of the magnet being passed through the centre of the wire coil in the experiment of 'changing directions within a coil'.

The voltage of an alternating current circuit is always changing or alternating up and down through the zero axis as a line on the graph. The line travels or flows in the shape of two hills. Once in the positive-volt zone and once in the negative, therefore, this machine is referred to as an 'alternator'.

The difference between the direction of DC flow and the polarity of the output of the windings of the generator is determined by the direction that the wire coil is wound, which is passing the pole face at the time of generation. During DC power generation in a generator, the poles are opposites, but the output is always above the zero axis, with a ripple effect as the top or the peak of the wave. The ripple effect at the top of the DC output is similar to a number of smaller hills, but all of the voltage is in the positive side, above the zero axis, whereas the sine wave (above and below the zero axis) is an alternating current.

The ripple flows along the top of the DC output as a very bumpy output wave due to the wire coil continually passing the three strongest lines and also passing all the other weaker lines. In a generator machine, this ripple wave (as the output of the windings) is then passed through an electronic circuit to flatten the ripple into a straighter output line of DC electricity.

Just quickly, a capacitor is the primary device of this electronic circuit which has the 'capacity' to store electricity and flatten the ripple. When a capacitor is placed across the output of the windings, it will clip the top of the ripple wave (of the output of the generator) to an almost flat DC line which may still have an ever-so-slight ripple wave, depending on the quality of the electronic output device of a generator.

The quality of the electronic device and the price of its components used in the design of the devices above will usually determine the DC quality of the output. Another common electronic device today is an electronic invertor, which can produce an AC output from a DC supply (battery supply turned into an AC supply). The invertor output is usually a 'square sine wave' or a 'pure sine wave' output dependent again on quality or price, plus application.

Eddy Currents

In the electrical industry, there are always extensive problems within every electrical circuit with the nuisance effect of eddy currents. These nuisance eddy currents occur within the electromagnetic energy, and they are generated within every electrical industry from power station alternator windings and anywhere else throughout the distribution network, plus the consumer wiring (houses), but will occur right down to the tiny printed circuit paths on printed circuit boards of electronics. It is vitally important later in this book to know that eddy currents do occur in the extremely weak energy of electronics circuitry.

Nuisance eddy currents in electrical circuitry are exactly as the name implies. They are vortexes of energy generated in large- to minute-sized eddies or gyres or vortexes of spinning energy within the electromagnetic energy which can and do cause excessive damage to every type of electrical circuit or equipment. It is also very important to know that very little is known about eddy currents except that they cause extreme mechanical instability in every electrical circuit and can or will cause damage in major to minor circuit components, plus the fact that they are very unpredictable.

I'll soon show you the details explaining that the energy of these four magnetic lines of the diagram generates eddy currents as tornadoes or twisters. The energy lines of neighbouring solar systems cause cyclones, hurricanes, and typhoons as we orbit through the lines in the identical polarities of **our Earth, plus each solar system line as they meet on the surface or crust.** All this will be very simply described in the details of slow reveals soon <u>about the way in which eddy currents within the electromagnetic energy of the lines generate the atmospheric vortexes,</u> but for now, I must point out some of the fundamentals about these atmospheric vortexes.

Tornadoes or twisters primarily occur above the equator in the northern hemisphere and only rotate in one direction. In the early years of reporting by the media, it was always tornadoes, hurricanes, and typhoons which were reported above the equator, and cyclones were recorded below the equator, but today the references sometimes appear to be mixed depending on the journalist. An important fact to clearly understand is that if an atmospheric vortex is being generated above the equator, in the northern hemisphere, then the rotation direction is permanently in a counterclockwise direction, no matter what it is referred to in today's language. Cyclones are much larger than a tornado, are more apparent below the equator, and rotate in the opposite or clockwise direction in the southern hemisphere vortex.

The official designation today is that a hurricane, typhoon, and tornado occur above the equator. Their identification is dependent on which ocean or which area of the Earth it occurs. Tornadoes are often referred to as twisters in America. A tornado/twister in America occurs in a very high percentage on land and stretches from the cloud to the ground. Tornadoes or twisters possess a much smaller or tighter circumference of the funnel, and they occur (in an extremely high percentage) above the equator.

But no matter which vortex, a tornado, cyclone, hurricane, or typhoon—either above or below the equator—will always rotate in the same direction as every other vortex or gyre in the same hemisphere. *For example, whether it is identified as a tornado, typhoon, or hurricane—which all occur above the equator—if it is an atmospheric vortex occurring above the equator, it will only rotate in one direction, which is counterclockwise.*

Science will tell you that the direction and motion of any gyre in a circular path is generated by the Coriolis effect, but I am describing the atmospheric circular motion as a mechanism caused by eddy currents. This theory is very new and must be considered strongly, knowing that eddy currents can and will occur in the very weak energy of electronics.

Under the laws of rotation for MY atmospheric eddy current vortexes, it says that they are generated under the same laws applicable to the direction of current flow within a wire coil when the magnet passes through the coil. *In support of my theory, let me point out that electrons were discovered 150 years before the discovery of electricity. This is because you could see the electrons under a microscope, but electricity is invisible except for lightning strikes. I identify these atmospheric eddy currents caused by the solar system lines in all cases as the **magnetic atmospheric vortex effect**. These magnetic atmospheric eddy current vortexes (on our Earth) are caused by the open-circuit king tide lines and also by the neap tide lines, plus at times other open-circuit lines. It is also very important to understand that these atmospheric vortexes do not occur in the period between the king tide lines of the diagram or in the period between two lines of the diagram.* In a later chapter, you will very easily understand how all this is possible.

Looking into the link below, you will see that the reoccurring dates and locations of hurricanes or tornadoes in Hawaii quite clearly appear during king and neap tides (with an occasional slight variation), suggesting that the four open-circuit magnetic lines of the diagram, plus missing lines, are the cause. The diagram lines which are missing are also the cause of tornadoes/twisters, but I say that the eddy currents generated within the electromagnetic energy of the three strongest open-circuit Cox-factor lines are the cause of a very high percentage of atmospheric vortexes.

Remember that while mentioning the three strongest open-circuit lines, I also described the presence of other open-circuit lines on either side of the three strongest lines. All together, the three strongest lines, plus the other lines outside those three, are the cause of tornadoes/twisters. A tornado/twister in the Americas or northern hemisphere is an atmospheric vortex which has a very tight circumference compared with a hurricane or typhoon which occurs in the same hemisphere. The evidence is unbelievably substantial.

Tornadoes/twisters are caused by the sun lines which radiate from our sun, and all other vortexes are generated by the sun lines from neighbouring suns. (I'll give you the details soon.) Recently in the news, I saw a very powerful tornado/twister in America (dated 4 March 2019) in a location where the highest percentage of tornadoes/twisters occur or are generated. Around this time, the date of the largest tornado fell on the king tide day, but there were also other tornadoes before the king tide, plus a tornado after the king tide, in that same area of America.

The reoccurring dates of tornadoes/twisters or hurricanes in Hawaii fall on common dates recorded as far back as 1843. They are recorded in the link below during the past 150 years and have mostly occurred on the fifteenth and sixteenth of August. Every tornado/twister on the reoccurring date 15–16 August must have been very significant. My theory states that they could only have logically been generated by a king tide line or neap tide line.

Other tornadoes/twisters were generated on other dates throughout the past 150 years (plus today's dates) and are also generated by the neap tide lines, but they may also have been generated **by one of the other open circuit lines** on the foreside or aft side of the neap tide lines.

You might assume that if a tornado/twister, as an eddy current atmospheric vortex, was significantly strong enough to have been reported and occurred on the same date in the same relative location, then it may be highly likely that it was generated by a king tide line but was not always the case. See https://en.m.wikipedia.org/wiki/List_of_Hawaii_hurricanes. Again, if the link disappears, then search for Hawaii hurricanes from 1843 or 150 years ago. This link identifies them as hurricanes and not tornadoes, but I can assure you that they were tornadoes. The identification of these vortexes in the link is a prime example of the ambiguity given by the media for atmospheric vortexes.

Some of you may have also already figured out that the most common reoccurring dates, 15–16 August, are not quite king tide dates in accordance to today's records of our oceans. The king tide date may have varied slightly over the past 150 years because our Earth is orbiting through the diagram lines and Earth is moving throughout the 500-year cycle of a planet alignment, or it may also have something to do with the shift in magnetic north over the past 150 years. The fact that the king and neap tides have shifted slightly in their re-occurring dates could be a subject that science can explain over the last 150 years. *In short, it may be that the reoccurrence of king tide dates may have shifted slightly due to a number of other variables.*

In a later chapter, I'll give a complete description of the mechanism which generates tornadoes/twisters, plus cyclones, typhoons, and hurricanes as atmospheric eddy current vortexes, in a truly simple explanation; but for now, I'll tell you that every open-circuit line (including the other open-circuit lines on each side) which exists as a part of our solar system has the ability to generate a tornado/twister as our Earth orbits through an open-circuit line when it is reactive with one of our Earth's ley lines. Remember that eddy currents will occur in the very weak energy of electronics. We all need to help explore this theory in depth by recording dates of tornados and tides.

As stated, I assume that these reoccurring dates and reoccurring locations above were caused by the strongest king tide line reacting with the ley lines within our Earth's field in August of every year up to 150 years ago. Remember the important fact of science which describes the movement of magnetic north in a total of about six degrees during the last 150 years, logically describing the movement of the closed-circuit ley lines within our Earth which will move the occurring location of tornadoes, plus the movement of our solar system.

I read somewhere that tornadoes do not occur on the equator where the twelve-inch neutral magnetic zone exists. Tornadoes today don't always reoccur in the precise location 150 years ago, which is also due the ley lines moving or relocating within our Earth due to the shift of magnetic north. There is also another cause for the slight difference in reoccurring locations of the tornadoes/twisters which are generated today.

I have mentioned the incredible length of our sun's solar system lines across approximately fifteen billion kilometres of our solar system. These lines are curved due to the resistive force on the front face of each line, and each line is a different length which will also cause a different degree of the curve or deflection that will apply to each line across that distance, and the lines are still growing. With a different degree of curving, that is deflection, this means that *a different length, thus a different curve, plus movement of the ley lines, will cause a different location on our Earth where the precise moment of reactivity will occur at just that peak moment of intense energy.*

When the eddy current atmospheric vortexes occur, they are reacting with a ley line of our Earth, and both of these electromagnetic lines must be at a level of reactivity for a vortex to be generated. I will describe this again in a later chapter after we get to know these magnetic lines a little better soon in the slow reveals.

A tornado/twister has a tight vortex in circumference compared with a typhoon, hurricane, or cyclone, which has a circumference that is enormous in comparison. I will also describe the difference in size between tornadoes, cyclones, typhoons, and hurricanes in those later chapters.

Cyclones reoccur every twelve months in the relatively same location and size, and there is a very logical reason for it. It is of utmost importance to know that cyclones, hurricanes, typhoons, and tornadoes/twisters occurring as eddy currents (briefly outlined above) in our north and our south hemispheres in opposite polarities also rotate in opposite directions **due to the opposite magnetic polarities of both the lines and of our Earth's north and south hemispheres**. These differing directions of rotation of water and wind in nature is referred to by science as the Coriolis effect. Again, I say that the cause of the different directions is the same law described above which determines the direction of current flow in the wire coil of the generator/alternator, *Fleming's right-hand rule*. More interesting and logical facts on this later.

As previously mentioned, every sun and its solar system in our galaxy, the Milky Way, has four lines which exist in the same formation as the diagram, all 400 billion of them (and growing). Each sun's lines also take four years to complete one rotation, but remember that we are also orbiting around our sun, and we orbit through our lines every three months or every quarter. Science says that the size of neighbouring suns are enormous. NASA says that one of our neighbouring suns is 1,000 times bigger than ours. Therefore, the lines of neighbouring suns are enormous in comparison to our sun. With the neighbouring suns rotating every four years, this means that the lines of these neighbouring suns will be pointing towards our Earth every twelve months, which is when cyclones occur, but I will give you more information and evidence of this soon.

In consideration of our Earth in the solar system, it sits (more or less) "out in the open" and is susceptible to the projection or injection of those neighbouring sun lines, except when our Earth is behind our sun or when the neighbouring lines are blocked by one of the planets in our solar system. It is important here to point out that tornadoes/twisters never occur at night, and this is due to the lines from our sun never pointing at that side of our Earth during the night. The neighbouring suns are in the northern hemisphere of our solar system, whilst our sun is below us. Remember that our solar system did align in the northern hemisphere during this time, placing our Earth between our sun and the neighbouring sun only once during each orbit.

The large atmospheric vortexes (hurricanes, typhoons, and cyclones) happen when the neighbouring sun lines are pointing at our Earth during the night and do continue to spin during the day under momentum due to the size of the vortexes. I will give you lots more information which will help understand this point about the larger vortexes.

I must also throw into the equation the fact that neighbouring suns' lines may also be blocked by planets within their own neighbouring systems. The fact that neighbouring solar systems contain planets has remained undiscovered until recently when an extremely large planet was discovered in 2018 at the edge of our solar system. It was recorded to be around 1,000 times bigger than our Earth and it was about time, wasn't it?

You may also notice that the locations of cyclones occurring in the southern hemisphere (which are much, much larger than tornadoes) do differ on our Earth. A high percentage of cyclones reoccur on the east coast and west coast of Australia; plus, cyclones also reoccur on the east coast of Africa, but all these locations are on the same longitudinal or horizontal path to one another on our Earth. Note that all these cyclones are also rotating in the same direction at each location because they are all occurring in the southern hemisphere.

The differing locations are due to two evidential facts:

1.   The neighbouring sun's lines are much longer in comparison to ours, suffering greater deflection and curving.

2. The neighbouring lines also exist in pairs by size, with the longer lines suffering even greater deflection, thus greater differences in the locations between cyclones, but each one is on longitudinal lines or paths.

(I'll give you all the particulars of the different types of atmospheric vortexes on our Earth caused by electromagnetic eddy currents generated within both 'our lines' and 'neighbouring lines' soon in those slow reveals in later chapters).

In the following chapters, I'll describe the process of a mechanism which generated and then created each planet, plus another holy grail theory describing why each planet exists in an individual colour, making up just part of the visible light spectrum, with the early science colours recorded as blue for Earth, red for Mars, yellow for Jupiter, orange for Saturn, amethyst for Uranus, and purple for Neptune. Amethyst is sometimes described as light blue.

The following chapter is a discovery that I made about rainbows which is in total opposition to the mechanism handed up by science about rainbows. It initially began for me as a simple hunch that I had about rainbows. This hunch was supported by other natural events that I saw happening on our Earth. This revolved around the fact that natural disaster events had a numerical sequence to them. I felt it necessary to tell you about my interesting observation of rainbows now rather than later so as to keep my reader interested.

<u>Rainbows and How They're Generated</u>

A rainbow, as we know, is a display of the seven colours of the visible spectrum. When you shine a beam of white light through a glass prism, it will display a light spectrum or a rainbow out of the other side, but it's not the only way that a rainbow is produced. In fact, nearly 400 years ago, Isaac Newton first proved that a rainbow can be produced by shining a beam of sunlight through a glass prism. He also took the rainbow which was projected out of the prism and shone it back through another prism to produce a beam of white light out of the other side. Let me state the very obvious here and point out that it had to be a prism-shaped glass through which he shone the light to produce this particular design of rainbow, better known as a spectrum of light.

Science describes a rainbow as 'refactored light shining through raindrops', but if the beam of sunlight were shining through each raindrop to produce a rainbow, then wouldn't it produce thousands of little rainbows due to it being projected from out of each raindrop? I believe that raindrops are involved in the projection of the colours of a rainbow but only to display the colours. The raindrops are only (more or less) a white screen for the rainbow to be projected onto. Let me show you my evidence and try to describe what is truly occurring when a rainbow appears in the sky.

My theory is that a rainbow occurs when a magnetic ley line of our Earth is pushed up or bubbled up into the air and when the ley line curves the magnetic energy to produce what I refer to as a refractured magnetic line. The ley line is pushed up or bubbles up into the air by the repelling energy of another magnetic source which is pushing the lay line, causing it to bubble up. Let me describe that other source.

The magnetic sun line from a neighbouring sun on the night side of our Earth is the source which pushes through our Earth to push a ley line up into the air. The neighbouring magnetic sun line projects all the way through our Earth to push the ley line up into the air. I describe a refractured magnetic line as a magnetic line which has been curved to display the seven colours of the visible spectrum, and also, another stronger magnet will cause a weaker magnetic line to curve or deflect.

I'm sure that you will agree that not every downpour of rain or every spray of a water from a hose or every waterfall in nature will produce a rainbow. In the following link to a number of internet videos, you will notice a truck in the first video driving along, spraying a fine mist of water from out of the rear of the truck. This fine mist of water is wavering (up and down) as the truck drives along, with a rainbow forming in a curve behind it, but the rainbow is very still or rigid behind the truck amongst the wavering water droplets as the truck travels along.

The rainbow (in this first truck video) forms because, at that very same moment, a magnetic ley line of our Earth is being simultaneously pushed up by another magnetic line from a neighbouring sun which has pushed through our Earth. (This might sound very wild or out there, but it is true). You will notice that the rainbow is in line with the path of the truck, but most importantly, the rainbow forms as the spray of water comes out of the truck. In accordance to my description, the magnetic ley line causing the rainbow is already there by coincidence, in line with the path of the truck. The rainbow is already there after the ley line has been pushed up, and the spray of water is (more or less) only revealing the colours which are displayed by each water droplet. The truck in the first video is spraying the water out of the back of the truck; the water spray is wavering up and down, but the rainbow is in one continuous, steady arc behind the truck as if the rainbow was occurring within a solid object in the air.

The second video is a greater example of the rainbow staying still or being rigid whilst the spray of water wavers up and down. The rainbow is undeniably already there in the air, and the water is simply revealing it. When you let go of everything you've been told by science about rainbows and watch the video over and again with my theory applied (on either of these trucks), it gives strong evidence that something is not quite right about the actual explanation given by science. If each rainbow had been formed by the water droplets, then the rainbow would also be wavering up and down along with the water spray. I'll show you some more solid evidence which backs my theory in just a moment. The same can be said for a rainbow that forms within the spray of a garden hose or a fire hose which is also wavering up and down whilst being sprayed.

This is the link to the videos of the trucks and rainbows: https://youtu.be/HV_uPG4tv8M. If this is unavailable, you can simply search for 'video of a rainbow behind a truck'. There are a number of them available.

In the link to the social media videos, you will also find a still photo of two irrigation trucks spraying out water onto a field. These two sprays of water are joined but display just one perfectly in-line complete rainbow. This is also very good evidence that a rainbow is a ley line of our Earth being pushed up out of the ground, and the magnetic energy is refracturing into the atmosphere. The magnetic ley line is bubbling up out of the ground.

But wait, that solid evidence is coming. In fact, my theory on rainbows shown by these truck videos is very much alike my other theory on magnetic lines being flat, in as much as our minds will not let go of the original theory handed up by science which has burnt an image into our young minds. A rainbow is NOT refracted light through raindrops. We have been told all our lives that a rainbow is refracted light displayed by the water droplets, and if we watch the truck video a number of times, we will see my theory of the refractured magnetic line slowly appear more and more within our minds.

But that's not all the evidence I have. Let me show you other strong supporting facts about rainbows which substantiate my theory of a magnetic ley line being pushed up and bubbling out of our Earth.

Before we continue here on rainbows, let me also gently remind you about the most amazing man in magnetics who ever lived, Mr

Nikola Tesla. He also had a great number of theories which were unseen to the naked eye when he told us about magnetics and also shunned by the science community, but the general public today now give great thanks to him for his hard work in magnetics. Please enjoy the following evidence about the true mechanism which produces rainbows, along with the video already mentioned above.

Why is it that a rainbow appears to have structure or to be almost solid in an innate form? You will see that in the videos. Here's that further evidence. We have all seen that a rainbow appears to be displayed (on average) about 300 to 500 metres above our Earth in the sky. For just a moment, imagine that a rainbow is produced by the mechanism I described above: the neighbouring sun line pushes up a ley line of our Earth out of its surface and is bubbled up into the sky.

Good evidence supporting my theory is that rainbows are never displayed in any part of the deep ocean which is deeper than about 500 metres. Ask any sea captain. This is obviously because the water is too deep for the rainbow to ever appear above the surface of the water after the magnetic line is pushed up through the bottom of a deep ocean. The rainbow is only ever about 300-500 metres above the solid surface of our Earth when it appears on land, and in the open ocean which is deeper than 500 metres (from the top of the water to the bottom of the ocean), a rainbow has NEVER occurred.

The continental shelf drops off almost like a cliff in its position in the ocean, and there will never be a chance of a rainbow after the continental shelf. Again, ask any sea captain who ventures into open waters deeper than, say, 500 metres. This information may be very true, but it's not actually the solid evidence I have promised you. Please read the following, which may very well astound you.

It begins with the fact that the sun lines are extremely synchronised in the way that they inject into our Earth. There is a rhythm to their timely events. This fact is best told in the timeline of events from where I made the discovery. Yes, folks, another story.

I was just leaving the local shopping centre one day when a storm floated by. With the storm came an incredibly huge double rainbow. (We know that there are sometimes triple rainbows.) I took a photo of it, and the information on the photo gave the timestamp. I had this extremely strong feeling that rainbows were somehow mechanical and that other synchronised rainbows were somehow going to be formed or generated by the magnetic sun lines, and so I vigilantly kept a lookout for other weather storms and rainbows in my area. I was looking for another timestamp.

It was by incredible timely coincidence that I was also in the area for another photo of another huge double rainbow just on the other side of town here in Beenleigh. This was just five kilometres south-west of the first. By calculating the timestamp on both of these photos, it told me that the two double rainbows were twenty-two days, three hours, and six minutes apart in that south-westerly direction by just five kilometres.

I didn't want to miss the next rainbow, which I felt was going to also be in the expected time frame as the other two, and so I advertised on social media for someone who could show me a photo with a timestamp of a huge double rainbow somewhere in the area around the time of the next prediction, but I didn't tell them a specific time. I simply said 'today or tomorrow'. A woman who was in a nearby suburb sent me a photo with a timestamp of a huge double rainbow, which was twenty-two days, three hours, and just two minutes away from the second huge double rainbow, plus five kilometres in the same direction. The three rainbows were in a straight line five kilometres apart by my estimates.

As I said, I did not tell anyone in social media that I was looking for a huge double rainbow to occur at any specific day or time. I simply asked for all rainbow photos to be sent to me with timestamps. There was just four minutes' difference in the third rainbow, and not only that but the three rainbows also were in a perfect line headed south-west, and each was approximately five kilometres apart.

When considering the ley lines of our Earth, we must remember that all are continually on the move due to magnetic north always being on the move. Because lay lines are always moving, then this explains why a rainbow appears to be moving whilst projected into the sky and not stationary.

Also taking into consideration that our Earth is tilted on an axis, you may also be able to see that a rainbow cannot form near the North and South Poles due to the angle that our Earth's surface is at. The aurora borealis is a distorted rainbow with a slight difference. It is a distorted magnetic line from our sun or from a neighbouring sun which is skimming through the sky of our Earth and being distorted by the energy within our Earth, causing the line to refract all over the sky.

I must also add that during a rainbow phenomenon, the rainbow will have a tendency to disappear as you approach it. This is largely because the rainbow is really very thin and can hardly be seen as you approach it from underneath; plus, of course, the rain and the ley line are also on the move.

On a boat in the Gold Coast recently here in Australia, I noticed that a very small rainbow appeared at the bow of this large aluminium boat. This was due to the magnetic energy within the metal of the boat curving out from the boat's bow.

Recently, days before sending my book to the publisher, a fellow who understood my theory on rainbows asked me if I had ever seen a rainbow on top of a huge mountain. You can imagine (with my theory above in mind) that a ley line would have to be pushed up through the Earth on a mountain before you would ever see any sign of a rainbow on top of the mountain. I haven't completed my studies here on rainbows, but so far, I haven't been able to find a photo of one on top of a huge mountain. There are plenty of photos of rainbows in the snow and other extremely cold regions of our planet, but there aren't any on top of huge mountains.

In closing, I would like to also add that I've followed lots of trucks and cars on the highway whilst it was raining or whilst the roads were wet, but I've never seen a rainbow travelling along behind the truck or car forming in the spray.

Moonbows

Rainbows are generated when a neighbouring sun line projects through our Earth in the same polarity to the hemisphere in which the rainbow is formed which pushes a ley line (of our Earth) up out of the crust. This repelling energy causes the magnetic ley line to bubble up into the sky and refracture, displaying the seven colours. As mentioned, the neighbouring magnetic line which pushes through our Earth must be in the same repelling polarity. For example, the neighbouring north magnetic line must be pushing through our Earth into the northern hemisphere and the opposite for the southern hemisphere. Please also note that the word refracture is an identification that I have made on curving magnetic lines (more on that later).

With the magnetic lines of our sun (in the diagram) in mind, my theory also logically said that a rainbow should also form at night, when our sun's magnetic lines push up a ley line out of the Earth on the opposite side of our Earth to our sun, and that is actually the case with a moonbow. The magnetic sun lines in the diagram are much stronger than those of neighbouring suns. The diagram lines make strong rainbows which only occur at night and are referred to as 'moonbows'. Moonbows are a great deal stronger than a rainbow, and as I stated, our sun lines are a great deal stronger than neighbouring ones.

I spoke to the captain of a boat here on the Gold Coast recently, who told me that she had been lucky enough to see a moonbow on land at night. It occurred on land when the moon was at its fullest, suggesting a king tide would have also been present in the oceans. This lady sea captain was also like many of the other sea captains I have spoken to and agreed that she could also never recall ever seeing a rainbow on the open ocean. I have asked quite a few oceanic captains if they have ever seen a rainbow on the open ocean, and they've all said no. A few of them have asked me, 'Why is that?'

The neighbouring sun lines which produce daily rainbows are not as powerful in magnetic energy as a moonbow. This is because a moonbow is generated by the magnetic sun lines from our sun, which are a great deal more powerful today than neighbouring sun lines. The sea captain of this Gold Coast boat also said that the moonbow was extremely dazzling, which is also an indicator that the moonbow would have been a stronger bow than a normal rainbow, and this is because our sun's magnetic lines would have a stronger magnetic energy than neighbouring suns. If the neighbouring suns' lines were just as strong, then the neighbouring lines would also cause or generate huge king tides.

When a rainbow occurs, the magnetic energy of a ley line is pushed up into the sky to a level of around 300 to 500 metres above our Earth when the line refractures in the sky. The sun's light (in a rainbow) amplifies the colours contained within the rainbow throughout the water droplets and displays it in the sky.

When a moonbow occurs, the same effect happens as a rainbow, but the light of the full moon is the source which displays the seven colours, although the sea captain said the moonbow was more of a shimmery grey colour. The grey colour of a moonbow is caused by the grey light of a moonbeam. The moonbeams which are highlighting the moonbow are grey because the light of a moonbeam is sunlight which has bounced back off the surface of the grey moon.

When a photographer uses a bright, shiny golden disk to reflect the sunlight onto his subject's face, the golden light will highlight the subject with a slightly golden tinge as though the person has a good overall tan. The evidence which supports this grey light theory can also be seen by the red colour of a sunset when the sun is shining through the red ashes after a bushfire. The sunset is red due to the colour of the ashes through which it is shining.

Before I finish here on the topic of rainbows, let me reiterate one important statement. I truly feel that rainbows appear in sequences of threes and that the sequence or timing that I've measured between the rainbows is accurate to about 0.0001%. Who knows? There could have been other rainbows which also happened in the expected time along with the three that I saw. The other rainbows could be single bows or even triple bows. Once I had three rainbows that were about the same time apart, I never bothered to look for any more. We need to involve social media a lot more.

Recently, a friend told me about their holiday on the beach in Hawaii, where they lay on the beach and observed a rainbow above in a circle which is nicknamed a 'sun dog', and I can only suggest that there was a huge magnetic rock or asteroid in the ground somewhere below where they lay that was reacting like a huge spherical magnet. In closing, I'd like to point out that there are many variations of a rainbow. There are rainbows which are double, triple, and on a rare occasion quadruple, but there are also circular rainbows and those that are inverted or smiley rainbows, plus rainbows which are straight across the sky.

My chapter here on rainbows are just an outline of your average rainbow. I must also add that, in a triple rainbow, this is an indication that the ley lines are positioned on top of one another because a triple rainbow is three ley lines on top of one another.

People on social media like to post photos of rainbows. The more that I pay attention to those rainbows, the more I realise that they are seasonal or come in groups, which ties in perfectly with my theory. I also like the idea of a 'pot of gold', but it just won't happen.

If I have rediscovered the mechanism which generates a rainbow, then I am entitled to rename it. I would like all rainbows to be known specifically as a 'Max rainbow'. I must also point out that Isaac Newton may have looked like he was bending white light, but in fact, he must have been bending a magnetic line of energy, and this is why I say that light energy and magnetic energy are directly related. Max is the name of my faithful K9 companion here.

How Our Sun Was Formed

The following chapter begins an outlined description of how our sun was first created, which really is quite simple, beginning with my contemplation of why the lines were curved. It's followed by a very logical explanation of the domino effect which created every sun and its solar system in our galaxy, the Milky Way, today. THE DOMINO EFFECT IS A MECHANISM WHICH CREATED ALL 400 BILLION SUN-STARS IN OUR GALAXY.

My Path to Discovery of How Our Sun Was First Formed

Part 1. The Reverse Engineering of Our Sun

The following describes precisely how our sun was first created across our galaxy, which will outline the effect of the domino's mechanism. This discovery is best described as a short story.

Some years after first discovering the diagram with its four lines in the giant science atlas, I began contemplating the cause of each line to be curved. It began when I simply asked myself a question which very quickly led to a number of overwhelming facts and Cox theories about our sun and its entire solar system, which I'll describe in very simple but logical details.

First, I need to show you a few need-to-know facts about the mechanical properties of magnetics. The four lines of our sun, as we now know, **span about fifteen billion kilometres** across our solar system to its boundaries, and just that fact alone helps us understand **why these lines do curve**. But to truly comprehend and understand curving of magnetic lines, we first need to understand that the curving occurs in stages, and I'll soon compare the curving to a bamboo cane as it is whipped through the air.

So let's take a quick look at just curving in general of other inanimate objects in comparison to curving of these four magnetic lines, but again, this is nothing too complicated. The following will genuinely help you understand that these invisible but very materialistic magnetic lines do possess mechanical or physical properties which cause the curving.

It begins again with that childhood experiment when we placed a magnet under the piece of paper and sprinkled iron filings on top. This revealed to us **the** most important fact of magnetics, which is magnetic energy exists in lines which also have mechanical properties; and although not as practical as a solid object, magnetic energy lines do display physical properties proved by the evidence given in the physical attachment of iron filings. (Simple so far, right?)

We need to now explore the fact that magnetic energy lines also have further mechanical properties so that we can completely understand that magnetic lines can actually bend and curve. To confirm that **bending or curving is possible** for magnetic energy lines of the diagram's size, length, and circumference, I first need to just remind you of the simplicity of another experiment that we also performed in our early school days with magnetics. When we placed the magnet under the piece of paper, we then placed another stronger magnet alongside it under the paper; and the original weaker magnetic lines (displayed by iron filings) began to deflect, bend, and curve. From this, we **can** clearly see that magnetic energy lines **will, to some degree, bend and curve**, especially when subjected to the mechanism of **magnetic deflection**.

Later in this book, I will explain that it was the curving of these four magnetic lines occurring billions of years ago which generated what I refer to as the 'planetary fracture points' within the energy of these four lines. Each planetary fracture point existed within a frequency and possessed what I describe later as 'magnetic resonance' when, eventually, in turn, each fracture began attracting particles as it began forming an immature planet after billions of years and then becoming a spherical mass. Let's now take a quick look at some similar facts about other innate objects which also curve so as to compare.

The Curving of a Bamboo Cane Being Struck through the Air

This is a very basic observation of a bamboo cane being struck through the air which will curve in stages and is a very good comparison of how curving of the four lines also occurred in stages but, most importantly, where the curving first began. When the bamboo cane was struck through the air, the velocity of the cane increased along with that of the hand, with the highest velocity reached at the very tip of the cane due to the arc. As the cane velocity increased, so too does the resistive force against the front face of the cane, with the greatest mechanical force causing a stress, placed on the fixed point, at the hand. As the forces at the hand increases, so too does the velocity of the cane and the force on the front face of the cane as resistance also increases.

Thus, curving will begin at the fixed point (in this case, at the hand) as the 'first stage'. It was also the case for each of the four magnetic lines. With the increasing velocity of the cane, followed by an increasing resistive force against the front face when curving will continue in a linear way (meaning it will begin where the greatest strain is placed on the cane at the fixed point of the hand) and travel out along the length of the cane in a continual pattern until the tip at the end of the cane. It is the same for each of the four magnetic lines. Curving of the cane is continuous or linear, which is what also occurred to each of the four magnetic lines but in stages due to the mechanical or physical properties of magnetic lines.

Now I am here to tell you that magnetic energy lines exist in the form of layers or tubes, which I will explain with strong evidence soon; but before we go any further, let me tell you about a short video I saw in social media. That video was made by a Scotsman who took a very strong magnet, placed it into a glass of water, and froze it overnight. In the morning, the frozen water revealed very tiny frozen tubes of ice extending from the pole face of the magnet. When dealing with the field of a strong magnet, any type of material can become affected by the magnetic field, including wood, but I will cover this fact in depth a little later. For more information about strong magnets and which materials are affected by them, just head to social media and watch some of the videos there.

In regard to curving of both the cane and each of our solar system's four magnetic lines, I must also point out that the longer and larger the circumference of a cane or a magnetic line is, the greater the resistive force will be on the front face which is generated due to a greater surface area of the front face. (Just logic so far, yeah?)

I mentioned earlier that each of the four magnetic lines exists in pairs or 'pairs of lines by size' and that different-sized lines radiating from our sun generated different curving forces to each pair of the lines. In the earlier years of our sun the lines were unmatched or in greater differing sizes. The two longer and larger pair of lines began to curve first (due to the force being generated sooner) before the other two shorter and smaller pair of lines curved. That one pair of lines began to curve before the other pair of lines is extremely important later when I describe how each planet was formed on each line within each of the fracture points. Briefly, I can tell you now that as the curving occurred in a linear way, so too did the fracture points along each line, resulting in the formation of a 'planet fracture' along each line, beginning closest to the fixed point, closest to the reactive centre sphere or ball.

Linear curving generated fracture points at linear stages along each line, resulting in the ascending order of the planets, which explains why the metallic planets are now closest to our sun today, eventually beginning their own orbit. The resistive force on the front face of the cane and each line does vary depending on the resistive value and the properties of the atmosphere where curving is occurring. There are generally two different atmospheres of both our Earth and space that will vary the curving.

Here on Earth, the resistive force is far greater than in space due to a greater density. Compared with the atmosphere in space (where a vacuum exists with less dust particles) for the four magnetic lines thus less force to generate curving in space but these magnetic lines are about fifteen billion kilometres long. The level of the resistive force on the front face will vary in each atmosphere. It is in a direct relationship to the composition of each atmosphere, which is proportionate to the density of each atmosphere.

So let's now make a brief comparison of the two different atmospheres, which is 'space under a vacuum' against the 'atmosphere here on our Earth'. But first, with regard to curving of the cane, the mechanical or molecular structure will also place a limit on the amount of curving which is mechanically possible before fracturing of the bamboo will occur. The molecular or mechanical bond between the atoms composing the bamboo is the governing factor that determines the eventual potential for fractures in the bamboo. This also applies to the four magnetic lines of energy, although not due to the molecular structure but due to the mechanical laws of the magnetic structure of the magnetic flux particles that has not yet actually been truly explored or uncovered.

The physical curving of each of the four magnetic diagram lines is directly related to the circumference and composite strength of each magnetic line and obviously due to a span of about fifteen billion kilometres across our solar system to our boundaries. How far, of course, any magnetic Line **can** also curve is dependent on the mechanical properties and composition or strength of the magnetic flux particles. There is actually very little known about the mechanical properties of magnetics, but so as to confirm the limits of curving, let's take a look at another mechanical structure of the molecules in the example of various timbers. By exploring the molecular limits on any tree, it may also help us truly comprehend that molecular properties apply to the physical and mechanical limits of everything and similarly to magnetic lines.

A Tree's Mechanical Properties

A Californian redwood is the tallest tree in the world, with a maximum recorded height of 116 metres, but the final height of a redwood is actually limited by the predetermined natural and scientific molecular structure of the species. The molecular structure of any tree will predetermine or limit its height and mass, and it, of course, will vary depending on the species and thus molecules.

Oversized growth places a stress on the stability of the molecules of any tree due to the specific molecule. This is a law which is applied to the building industry when it comes to choosing the timber for the span of any timber beam. (Please remember, this is written for those of younger age and older.) It's often said that the original energy or source contained within the seed of any tree is the final determining factor for the shape, health, and mass of that tree, a fact which is dependent on the laws of nature.

The size of a tree due to its molecular structure is just one example of the 'intelligence of a tree' with many other forms also in play. Intelligence is also seen in the type of fruit, the type of seed, and the manner in which it propagates. The tree's defence system is also intelligence by the way of thorns or sap; plus, the leaf design and a whole host of other factors contribute to a tree's intelligence. (The list of items on the intelligence of a tree is as long and varied as the number of species.)

*Just before we look at more exciting facts, I feel the need to make another personal statement in regard to **our trees**, which are the lungs of our Earth. I can tell you that, throughout my own work-life experience as a tradesman, recovery after a day of hard physical work is becoming harder to find due to our Earth's atmosphere and oxygen percentage, plus the pollutants. If you think that it has to do with aging, just look for one second at footballers in their youth who now use oxygen tents and decompression chambers after a game so as to find full recovery.*

It honestly has nothing to do with age. I suddenly became aware of it around the time that they began cutting down the Amazon forest in massive amounts each day to make way for cattle and palm oil production, where hundreds of square miles were devastated, which hasn't stopped. Please take the time to investigate and tune in with our Earth so as to find out for yourself. It's too late to measure just how important the oxygen production by the Amazon forest is.

*Take the time to also explore the ever-increasing number of **JUMBO JETS** being added to **OUR SKIES** throughout the entire world which consume the most incredible amount of oxygen being burnt each second within each engine and sometimes four engines on each jumbo. Please contemplate for just one minute of your life how much oxygen just **ONE***

JUMBO JET ENGINE BURNS EACH SECOND, *the number of engines per plane, the incredible and increasing amount of oxygen being consumed during take-off, and the insurmountable* ***NUMBER OF JUMBOs*** *taking off each minute, plus the ever-increasing number of flight hours, the* ***NUMBER OF FLIGHTS*** *each jumbo is taking and burning up oxygen for, and finally the number and SIZE of airports, which are also increasing globally. Do the math.*

The Values of Resistance in the Atmosphere of Space Compared with Here on Our Earth

So as to truly confirm that the four lines have physical or mechanical properties which are able to curve or bend, we need to consider a few more of the basic facts about mass, velocity, and resistive forces in different atmospheres, but it is nowhere near as complicated as it sounds.

If we contemplate the resistive force which will be placed on any molecular structure at a specified velocity, travelling through space, which is under a vacuum, compared with here on our Earth in a denser atmosphere, it will help us clearly see that Earth's atmosphere does have a greater resistive force. The force against mass as a resistive value increases as velocity increases but differs in each atmosphere. There will, of course, be an actual scientific formula for each atmosphere, but we're going to be looking at it today in lay terms.

Mass as we know is size and weight. Science says that everything does possess mass. The value of the resistive force is directly proportional to the mass, the velocity, the surface area, and the atmosphere it is traveling in, which is the basis of Einstein's formula for relativity. What I'm about to show you is that space has a lower value of resistance due to the vacuum than that of our Earth, and to help us understand this fact, let's take a quick look at the following.

If we were able to place a glass cylinder over the Eiffel Tower, all the air contained within that cylinder in our Earth's atmosphere would outweigh all the steel that the tower was constructed of. Our Earth's atmosphere appears surprisingly heavy for its value of mass, more than, I'm sure, what we were ever aware of. To make a proper comparison of these two atmospheres, we need to consider a short video I saw of astronauts moving about in space.

In the video, astronauts were seen swimming about in space while paddling with their hands, generating a resistive force against the front face of their hands, which enabled them to paddle about in the thinner atmosphere of the vacuum in space. If the atmosphere of space had no resistive values, then movement, of course, would not be possible for the astronauts.

Remember, science says that everything does possess mass, therefore suffering a resistive force, including space; and with that in mind, let's look at the four magnetic diagram lines.

The Speed of Light in Space Compared with Here on Our Earth

Science recently announced the irrefutable FACT that **light and electricity** are absolutely and positively **directly related**, stating that each element will travel at an **identical velocity when under a vacuum**, which of course is the atmosphere of space. My guess is that this was discovered when changing from electrical telecommunications transmission in copper cabling over to light transmission by way of the fibre optics cables. The discovery of a difference between the two velocities (of light and electricity) during transmission turned out to be much slower than expected after being explored in a laboratory, with light and electricity under a vacuum.

A fibre optics cable on Earth cannot, of course, hold a vacuum; and as it turns out, light transmission within the fibre optics cable is only about 100 times faster compared with electrical transmission within copper cabling, which is actually surprisingly slow or at least slower than expected.

Let's just look at that statement for a moment—identical speeds when UNDER a vacuum, and space is under a vacuum. The atmosphere that 'light' being transmitted within a fibre optics cable is, of course, our Earth's atmosphere. Light is said to travel by way of photon particles.

Just briefly with regard to the term 'particles', science says that the electrons, protons, and neutrons of an atom are atomic particles. There are also smaller particles also recently recorded on a TV program identified as subatomic. The reference to 'particles' can vary a great deal, from subatomic particles mentioned above (particles the size of dust particles or breadcrumbs) to everything in between. Personally, so that I can understand how a resistive force against a single photon or light particle is at all possible purely due to its mass and the surface area of the light particle's front face, I try to visualise photons as specks of dust travelling through the atmosphere. Just the sheer fact that a photon of light does vary in velocity due to varying atmospheres (generating different resistive values) caused by the physical properties of photons, I think, is truly amazing.

Here is a recap:

a. *The resistive force against the front face of the cane is the cause of curving, where the force against the front face of just one photon of light will also cause it to slow in velocity, which varies both here in our Earth's atmosphere and in space but has a greater resistance in our Earth's atmosphere.*

b. *There is an all-encompassing counteractive effect on a cane caused by the physical properties which causes the curve and also the four magnetic lines in space, generated by the same mechanism of the resistive force.*

c. *Light travels faster in a vacuum or in space; plus, it has a higher velocity, which causes the cane to curve, but has a lower value in space due to less resistance in a thinner atmosphere under a vacuum.*

d. *Most importantly, photons of light do possess mass and will suffer a resistive force on the front face of the photon due to the surface area, which will decrease the velocity and will vary in either atmosphere.*

Because magnetic energy can move solid objects and light has very little effect on any object, I'm sure that you'll also agree that magnetic energy which exists in lines (held together by magnetic flux particles which are flowing) will suffer a far greater resistive value on its mass than any one photon particle of light, but magnetic energy also exists alongside electrical energy functioning. However, a magnet exists in an extremely high percentage of magnetic energy; the same is true for the four lines radiating out from our sun in the diagram.

The bottom line here is that magnetic energy in lines, which are flowing in magnetic flux particles, will have far greater physical properties compared with light particles and therefore suffer a higher value for their resistive force. These facts are very basic but also very logical, helping us truly see that there is a resistive force against each of the four magnetic diagram lines at a higher value of resistance than the identical force against a single photon or light particle due to the magnetic energy possessing a greater value of physical mass.

For the next step in my slow reveals, we need to look at the following. Science says, 'Light and electricity are directly related,' but only after discovering that they both travel at identical speeds when under a vacuum. I believe the reason why light and electricity travel at identical velocities when under a vacuum is that, during the period in our galaxy before the birth of our sun, in a time before the very first atom was ever ignited, there wasn't actually any light when all space was dark, contrary to what we know today.

Our sun did exist before our Earth, and if light and electricity only travel at identical speeds whilst under a vacuum, which is the atmosphere in space, then **space was therefore the birthplace of light when the very first atom was ignited by a sun-star.** *I'd like to just politely ask science that if they announced the discovery of identical speeds between light and electricity (as a NEW discovery), then why did they NOT include magnetic energy? Magnetic energy and electrical energy function together as one, and isn't it electromagnetic energy? Therefore, magnetic energy will, of course, also be travelling at the identical velocity of electricity, or will it?*

This link has information on the electromagnetic spectrum and is designed for kids: http://www.physics4kids.com/files/light_emptye. html. *Please just read the first heading and then scroll down to the electromagnetic spectrum and someday have a further read for yourself of the entire link. If the link disappears, simply search 'electromagnetic spectrum and light waves'.*

Plus, I'd also like to just gently remind science that every energy on the electromagnetic spectrum (discovered by science in the last century) is described as 'waves of light energy' existing in frequencies which are both visible and invisible. Does that NOT also mean that every energy, including light and electrical energy (as electromagnetic), is ONE, therefore travelling as ONE at the same velocity as the speed of light in space? After all, every electromagnetic energy on the electromagnetic spectrum is a frequency of light energy travelling at the speed of light in space where a vacuum exists.

*Plus, the most astonishing and maybe somewhat forgotten fact when we can't see the trees for the wood is that science describes every frequency of light as electromagnetic energy travelling in the form of photon particles, which are actually **PARTICLES OF LIGHT**. So as far as it being a NEW discovery of a relationship between the speed of light and electricity, can I ask science, 'Are you absolutely certain it's new?'*

To properly understand the physical properties of the four magnetic lines which suffer a resistive force which will cause bending at a specific velocity through space, we must look just briefly at the two major facts about these four magnetic lines. The following is my last description of the simple mechanics about magnetic lines, explained in the most uncomplicated terms, under two different conditions.

The First Condition

The first is the example of the early childhood experiment shown to most of us in primary school. (I hate to do this again, but it is important.) The iron filings reveal the physical structure of magnetic energy in lines, but it also describes magnetic energy in the simplest way by showing it to us in the form of lines. Most importantly, when we place another stronger magnet underneath the paper in the same polarity, the incoming magnet will deflect the magnetic lines of the weaker magnet, displayed by the movement of lines in iron filings. The simple fact that the magnetic lines are deflected confirms that magnetic lines have elements of mechanical structure which are limited and will, **to some degree, curve** as they do during the magnetic deflection.

The Second Condition

Well, the second condition is actually extremely simple. These four magnetic lines are curving due to the resistive force, which is also possible in space despite the lesser resistance caused by the vacuum when resistance is generated on the front face.

The Reverse Engineering of Our Sun

Stage 1: What Would Our Sun Have Looked Like in Its Early Stages of Development?

Whilst I stared at the curve in each line, I began to contemplate the reason why they were curving. I theorised that this is due to an increasing reactivity within the reactive sphere which, in turn, increases rotation velocity, thus increasing the resistive force against the front face of the lines, causing the lines to curve.

The rotation velocity of our sun is directly related to the level of reactivity within the energy generated by the reactive sphere or ball. This occurs in a similar way to an electric motor which increases in rotation velocity where the number of poles will govern the final top speed or rotation velocity of the motor. The four poles at every sun will eventually govern its final rotation velocity, but I further theorise that the final top speed in rotation velocity of each sun has not yet been reached as they all ever so slowly increase in speed.

Top speed will not be far from the rotation velocity, which is maintained today but will only eventually be reached after trillions of years due to two main factors:

1. The incredibly long span of the lines across our entire solar system to the boundaries, recorded by science to be about fifteen billion kilometres
2. The final top level of reactivity not yet being reached by our sun, especially the sphere, which is the major factor, making it only similar to an electric motor.

However, the bending of the four lines within one sun-star will one day reach their limit when the lines will eventually overlap, but which sun-star in our galaxy will do this is the part that I'm not sure of.

Stage 2

I further contemplated that the reactive component of our sun is the mechanism which attracts atoms that are being ignited. In fact, every sun in our galaxy attracts the identical atoms of gas which ignite the instant that they arrive, and soon I'll tell you why identical atoms of gas are attracted to every sun.

With atoms continually being attracted and then burnt in an instant, I felt that the removal of atoms from the reactivity of the sphere would somehow wind back the clock, and I would reach a point where the rotation velocity would slow. Remember that our sun's reactivity and rotation velocity, thus resistive force, caused the curving of each line and that **an earlier stage or younger sun simply meant less reactivity**, hence slower rotation, thus less resistance and less curving.

As I began to visualise the removal of atoms, thus less reactivity, the rotation velocity began to slow, with each of the four lines logically also beginning to straighten further and further with less and less reactivity, but what would our sun have looked like if the reactivity continued to lessen? Would the lines have become totally straight?

Again, removing more and more atoms and visualising less reactivity, I began to see an even slower rotation, causing a reduction in the resistive force, arriving at a lesser curve, resulting in straighter and straighter lines which were eventually straightened as the clock was wound back. I realised during this 'stage of growth' of our sun (as it evolved), the four lines were once actually straight due to a period when there was lesser reactivity, which meant that the sphere would have also reduced in size.

I had arrived at stage 2 of the reverse engineering of our sun when our sun once existed with straight lines. **The stage 2 image of our**

sun had now appeared with straight lines, which opened a gigantic door to the next incredible and astonishing discoveries. Let's go now and see my next slow reveal, which unfolded after I simply began asking myself again, 'What would an even earlier or younger sun and its lines have looked like if I kept the clock going backwards?' Have you got your hat on? You're going to need it.

Stage 3: The Incredible Discovery of an Even Younger Sun

Looking into this new stage 2 diagram with straightened lines, I began to wonder again what an even earlier or younger sun would have looked like, and so I began the process of contemplation. The lines were straightened due to reduced reactivity within the sphere; nonetheless, a reactive sphere was still present and an earlier or younger sun simply meant less reactivity again, right?

At this point, I began asking, 'What would be taking place at this point within the reactive sphere?' And I began visualising atoms still being attracted, accumulated, and burnt due to the reactive component. Less reactivity clearly meant less attracted atoms, which reduced the reactivity and thus size of the sphere due to less atoms. I then began visualising the removal of more atoms, reducing the size of the sphere again, until I eventually arrived at an image of our sun with straightened lines and a relatively smaller reactive size of the sphere as stage 3.

Stage 4: The Sphere Slowly Reduced in Size

With the clock winding back as I removed more and more atoms, the reactive sphere continued to reduce in size, shrinking further and further as less and less atoms were present with less reactivity when a smaller reactive sphere appeared. At stage 3, the sphere was relatively small, but reactivity was still present, and I continued removing more and more atoms until the sphere was non-existent. This period in the life of our sun was at a time when just a few atoms lingered in the junction of four lines. This was the birth period of the reactive centre sphere.

At this point, I visualised that it was just a junction of four lines with no atoms, which meant a total absence of reactivity; but I quickly realised that, despite the lack of reactivity, the mechanism of attraction was still ever present as stage 4. The image of stage 4 was no atoms and no reactivity, but an ever-present component of attraction was present within the junction of four lines. *The reactivity of atoms within the sphere consists of a high percentage of hydrogen gas atoms (98%), which have always been the primary atom to be attracted, accumulated, and burnt at our sun. Helium is also present along with a number of other minor gases.*

This new stage 4 image of our sun with a junction of four lines and just a few hydrogen atoms lingering around the centre of the junction suddenly became very logical and real. I also realised at this point that, with just a few atoms lingering in the junction, this image was just a short moment and was the true moment of the birth of our sun.

Stage 5: The Master Key Was the Next Step in Discovery

When visualising the junction of four lines and an absence of atoms, I began to contemplate how it was at all possible for the existence of a junction of four lines, but more importantly, **where had the lines come from?** The next few steps lead to what I refer to as the **master key** which opened the door to exactly how the junction was created and precisely where the lines had come from, which of course described **how our sun was created**.

Let's take a quick look now at the short story of the events which led to the overwhelming discovery of how our sun first began. The story is really quite exciting and came in a sudden moment of inspiration that struck as hard and fast as a lightning bolt when I saw it. I believe it was a historical moment and one that we all would never forget. I hope it comes as exciting for you now as it did for me when I discovered it. Let's go. HOLD ON TO YOUR HAT, FOLKS.

During that split second that it took the lightning bolt to strike, I was actually just sitting, staring around the room, pondering the junction and lines in the new image of stage 4. I sat just staring at the image, which was our sun with no reactive sphere but lines crossing and totally void of atoms. The lines met within a fine point as the 'junction' possessing a very reactive mechanism. The reactive component also showed the attraction of atoms which existed in the identical frequency as the reactive component that generated the attraction of atoms which were primarily hydrogen atoms along with a small percentage of helium gas atoms amongst others.

After quite some days and long hours whilst I sat there staring at the junction, knowing I was close but not really comprehending what I was about to discover, I asked the same question over and over again. No atoms and four lines, but where had the lines come from? In my worst moment of frustration, I'm beginning to get angry after prolonged contemplation when I happened to glance across the house into my son's room, clearly visible from where I sat. Then the lightning bolt hit me, followed by the CLAP of loud thunder as it struck.

At the time, I was just staring aimlessly at a poster on my son's wall, a mathematical chart full of simple fractions and mathematical symbols. I was just sitting there, staring at just one particular symbol for absolutely no apparent reason. My mind was actually off in a daze when my subconscious tried to alert me. I was dreaming about my childhood during a night at our city's exhibition showgrounds. I was there watching the precision driving team as two cars raced towards each other's path.

I was almost hypnotised and daydreaming when—BANG!—THE BOLT OF LIGHTNING STRUCK. I was struck by a billion volts in the face like a brick. I was sitting there, staring at a mathematical PLUS symbol, watching the two race cars in my mind crossing paths, when I said to myself, A PLUS SIGN IS JUST TWO STRAIGHT LINES. I wasn't looking at four lines in the stage 4 diagram forming a junction, which I was supposed to be contemplating. It was just **TWO LINES THAT CROSSED OVER.**

Our sun first began from two magnetic lines which crossed over, generating a simple junction which obviously possessed magnetic resonance due to the attraction of atoms which also existed within the same frequency. I was in total AWE at what I'd just discovered. It was how our sun had first begun.

Stage 6: Had the Junction of Two Lines Generated Reactivity?

I contemplated the 'junction', which very obviously had an extremely reactive component within a fine centre point generating reactivity, which had also attracted, accumulated, and burnt hydrogen and helium atoms, but what was this mechanism identified in the science community? In my theory, I had identified it as 'magnetic resonance', but for it to possibly be magnetic resonance, I would have to find proof of its existence in the science world. In all my years in the electrical trade, I had never heard the term.

Magnetic Resonance

Every sun in our galaxy, the Milky Way, exists in this identical form, with two lines crossing over, which eventually became four, radiating from a reactive centre sphere out into a solar system, all 400 billion of them, with every one of them burning a high percentage of hydrogen gas atoms. Space is recorded as having an **extremely** high percentage of hydrogen gas atoms at around 97–98% with every single gas atom in transit, all being attracted, accumulated, and then burnt within a reactive centre mass of a reactive sun-star or sphere. The hydrogen gas atoms being burnt at our sun, plus every other sun, are all singular atoms, meaning there is no other atom bonded and therefore not a molecule of gas.

My son told me recently that, in regard to atoms, liquid atoms are described (in lay terms) as atoms which slide over one another; solid atoms, like those of timber and stone, reportedly jump around, but gas atoms are continually bumping or smashing into each other, depending on how volatile they are, resulting in friction to differing degrees, making some gas atoms potentially the most volatile or unstable. Hydrogen atoms are amongst the most unstable atoms known to man. Remember the Zeppelin.

Gases—like liquid petroleum gas (LPG), natural gas, or chlorine gas—are all **gas molecules consisting in a number of atoms which are bonded**. Molecules form together as compounds. I contemplated for quite some time that the junction of lines in the stage 4 diagram was attracting an extremely high percentage of hydrogen gas as singular atoms, and the junction, which existed within a frequency that also possessed resonance, also possessed the same frequency as hydrogen and helium together, but I'd never ever heard of the term 'magnetic resonance' except for the name of a medical imaging machine. (A molecule made up of a number of atoms is a little like a number of instruments playing together which make up a tune.)

I contemplated the basics that the four lines existed in a very high percentage of magnetic energy but in the same resonant frequency, which in turn attracted hydrogen and helium logically due to the magnetic resonance, but the problem didn't go away that easily. I'd never heard the term 'magnetic resonance' ever in all my years in the electrical trade.

I made further contemplation that magnetic energy and electrical energy function together as one, with electrical energy always expressed in a wave form as a frequency. Logic said to me that if they function together as one, then magnetic energy must also possess or exist in a frequency. **I began to contemplate resonance in every example I could think of.**

It was around this point when I began subcontract work in a government railway yard as an electrician where I had lunch in an engineer's lunchroom. The lunchroom was full of reading materials for the engineers. It had encyclopaedias, all sorts of technical books, and science magazines to stimulate the mind during their breaks. At the time, my reading skills weren't competent enough to pick up an encyclopaedia, but I did see an old book that caught my eye called a 'science dictionary'.

I looked up the word 'magnetic'. I scrolled down from the word 'magnet' and found the term 'magnetic pump'. The dictionary was very old and described a magnetic pump as 'magnetic lines that cross over and attract atoms of gas, heating them to 1,000,000 deg.' The description was straight to the point but made no mention of the mechanism which initiated the attraction of the atoms of gas.

I asked myself this question. 'Is magnetic resonance just in my mind, and does magnetic resonance exist?' But I thought about electromagnetic energy again; be it magnetic or electrical, it must obviously exist or possess a frequency. After all, the two energies are never apart; they're both electromagnetic energies. I contemplated the pulse which our Earth emits from the molten magnetic core. The lines are primarily magnetic energy which form a junction where the lines cross over into a point existing within an **individual frequency** and also emitting a frequency that possesses magnetic resonance, attracting any atom that exists within the same frequency. This was the basis of my theory on magnetic resonance. The junction attracted atoms which existed in the **identical frequency** into the fine point within the junction, and it had to have done so under the law of resonance, but was it magnetic resonance?

The subject of resonance is an extremely important subject throughout this book and must be properly understood to fully understand my other theories in later parts of this book, and so I need to show you resonance in other forms for a number of reasons. Let me describe a few examples of resonance but in quite detailed descriptions so as to completely explain that it is **magnetic resonance** which is occurring during the crossover of the lines. I spent quite some time contemplating magnetic resonance so that you too can fully comprehend the term. Let me give you these examples of other types of resonance in other modalities.

Swimming Pool Resonance

This is the simplest form of resonance and a very good example because most of us played with resonance as children in the swimming pool; plus, it is a visible example. Playing around in an above-ground pool as a child, I would get the boogie board and push it down in the water in the centre of the pool, making a wave that would bounce back off the outer wall of the circular pool. The wave would travel to the outer wall, bounce back off it, and travel back to the centre as a concentric wave in a circle. I would push and push with timing to increase the wave's height until the circle wave would collide in the centre and force a resultant fine spout up into the air as an amplified direction for the energy force. This occurred as the circle wave returned from the outer wall, all the time slowly closing in until it collided in the exact centre.

It is very important to clearly understand that from the moment the circle wave bounced back off the outer wall and began travelling back towards the centre of the pool, the concentric wave had begun closing in all the way until it collided within itself. The fine spout up into the air was a result of the concentric wave meeting in the absolute centre, having been amplified as a resultant collision when the wave energy was forced upwards as the direction of least resistance. The fact that the wave would meet and shoot upwards is a very good example of resonance, where the timing caused the concentric wave to meet with itself in the centre and where the resultant energy (as the force of the wave) would shoot the water up into the air in the form of a fine spout as it collided. All the water in the base of the pool meant that the direction of least resistance was upwards and not down.

Resonance in a Wine Glass Vibrating

This is one example I'm sure that we've all done. Take a wine glass and rub your wet finger around the top of the glass until it begins to SING. When your finger rubs the glass at a particular speed, the glass will begin to sing but not before or after that specific speed. The molecules in the glass become excited by the vibration, which is in the exact frequency of the glass vessel. If the glass is sounding at the exact resonant frequency of the glass, then the molecules will be unstable at this frequency. If you tap the inside of the glass whilst the molecules are racing and unstable, the side of the glass will simply fall out.

I know this is a fact because my brother talked me into performing the experiment on one of the expensive wine glasses my parents owned. You guessed it. I was in big trouble.

For the wine glass to start singing, you must rub your finger at just the right speed. What is actually happening is that your finger is jumping at just the right frequency of the glass.

Resonance in the Waving Suspension Bridge in Tacoma, USA

You have all seen the video of a suspension bridge waving in the wind during the last century in the city of Tacoma, USA. In the video, a huge cable suspension bridge is waving in the wind like a piece of paper due to a resonant wind energy. Wind exists in a physical wave form just like the water waves in the pool description above.

The bridge waved due to an incoming wind wave (generated by a gale force) which travelled through the bridge, bounced off the cliff behind, and travelled back to where it met with the incoming wind wave in the middle of the bridge, which just happened to be the weakest part. The recoiling wind wave bouncing back off the cliff needed to be in perfect resonance with the incoming wind wave as it returned through the bridge, meeting with the original incoming wind wave in the precise moment in the weakest point of the bridge. The bridge waved up and down due to the collision of the two wind waves, generating an amplified fine spout of wind energy upwards as they collided. This needed to be at the precise frequency for it to occur, meeting and amplifying the wind wave. When the two wind waves met with each other in the middle of the bridge, they initially went in two different directions; but because the water acted as a wall below the bridge, it was initially pushed upwards into the air as the direction of least resistance. Just as the water in the base of the pool forced the water spout upwards, so too did the river below with the bridge.

It's important to realise that the resultant collision of energy between the two wind waves generated amplification, and it was a very narrow force which was directed upwards due to the river being below. It was simply gravity that allowed the bridge to fall back down in the period between resonant waves, just as the circular wave of water in the pool had also fallen back down. The bridge is another very good example of resonance, this time in the form of a wind wave.

Resonance in Two Sounds, Which Will Generate a Mechanical Vibration due to the Presence of a Physical Wave

Resonance in sounds can occur, and when they do, they also generate a mechanical wind wave which also generates a secondary physical vibration. Let me describe the mechanism.

When sounds exist in resonance, a wind wave is generated by the physical vibration of the source, be it two speaker cones, two trumpet horns, or two strings on separate guitars or violins. As a musician, you may be able to tune any musical instrument by using the ear, but a stringed instrument can also be tuned using the mechanical vibration emitted from a tuning fork. Let me show you the mechanics.

We simply fold a piece of paper, place it over one string (let's say the A string of a violin), and then strike the tuning fork (in the same musical note of the string). Whilst the tuning fork is still vibrating and dispersing a mechanical wave, we then place the base of the tuning fork onto the timber body of the violin and adjust the string until the piece of paper vibrates. The string vibrates in turn, making the piece of paper vibrate whilst the violin string is tightened or loosened to eventually exist in the perfectly timed vibration of the fork when they resonate.

The physical evidence of two resonating sounds is displayed by the piece of paper when it physically vibrates, caused by the resonant mechanical wave which physically vibrates the wood and then the string, causing the piece of paper to vibrate, generated by the wave of the tuning fork. As stated, resonance is an extremely important factor here and must be understood completely. There are certain aspects later in the book which also incorporate resonance, and because of its importance, you must especially fully understand the following.

Whilst the tuning fork is mechanically vibrating, which is vibrating the timber of the violin, the string is being adjusted, and the amount of physical vibration generated in the string and paper will increase as the frequency of the string becomes closer to the frequency of the fork. As the string becomes in tune, it begins to resonate. The mechanical vibration emitted by the fork will physically vibrate the timber and thus the string and paper at its strongest when the two waves are amplified during perfect resonance, just like the pool and the bridge.

In every example so far, the amplification caused by resonance occurs due to the two waves melding or amalgamating, much like the double bounce on a trampoline. In this example of the violin's string vibration, the strongest vibration will occur during resonance, and a slightly weaker vibration of the string and paper will also be generated on each side of the strongest and perfectly resonant vibration. The slightly weaker vibration is present before and after perfect resonance, and this is due to the string being almost in tune or almost within the resonant frequency or when it is ever so slightly out of tune. The slightly weaker vibration on each side of perfect resonance can be clearly seen if you watch the piece of paper carefully.

The strongest vibration occurs during perfect resonance, but a slightly weaker vibration also exists on each side of that and can be directly assimilated to the three strongest magnetic lines or the 'triple dipole lines', and I'll show you stronger evidence of that fact soon. The fact about being almost in tune, which generates an almost in-tune vibration, is what must be clearly understood before reading further. Those slight variations in magnitude of the mechanical wave generate a slightly different vibration which occurs whilst adjusting the string, as stated, just before and after resonance of the string.

The slightly weaker vibration is generated by an almost in-tune frequency wave form, and the strongest vibration is present during perfect resonance. The slightly weaker vibration on each side of perfect resonance is due to the presence of a slightly different frequency, generating a slightly different wave on either side of the perfect resonant wave or the wave of the highest amplitude. The existence of these 'three vibrations' generated by three slightly different frequencies is a subject of utmost importance when reading the later chapter on volcanoes and mountains. I understand that three differing vibrations on a violin string whilst being tuned might sound a little peculiar in relation to volcanoes and mountains, but I assure you it will all be explained later. For now, please just take note of the existence of three vibrations generated as a result of three slightly different frequencies close to one another.

Electromagnetic energy is somehow directly related to sound reflected in the waves. Logic will have to also say that wind waves, plus water waves, are also somehow similarly related, but that will take a whole week of work or another book to explain it. We can, of course, also tune in a violin using an electronic tuner, but the fact of three slightly different vibrations generated by three slightly different frequencies above is extremely important—almost in resonance before being in tune, perfect resonance when in tune, and almost in resonance after being in tune. (These terminologies are about the best I can explain it.)

Two identical 'electrical' frequency waves generate a mechanical or physical vibration. I'm sure that we are all familiar with a transformer that makes a humming vibration and sound. This is due to the presence of two different electrical currents which share an identical frequency wave. They are both being transmitted at the same frequency. The waves are mechanically colliding, causing physical resonance vibrations to the metals or laminations of the transformer, including the windings. The frequency of the wave within the 'primary winding' of the transformer is meeting with that of the 'secondary winding', which exists in the same frequency of 50 Hz here in Australia.

A transformer is a coil of wire using the primary energy source of electromagnetics to generate an energetic magnetic field and then a collapsed energy field. It is the energising of the magnetic field and then the collapsing of the magnetic energy field, which in turn generates movement of magnetic flux particles, thus generating electron movement. The primary energy which is present within the mechanism of any transformer (from substations to electronic circuit boards) is magnetic energy, with a somewhat even level of electrical energy, but the specifics or the percentages of these two energies have never been actually calculated, and I'm merely estimating the primary energy as magnetic due to my own experience.

The motion of the waves in the magnetic component generates a mechanical or physical wave generated within the frequency of the magnetic energies. A basic example of magnetic energy, existing and possessing a greater secondary mechanical energy, is seen in a magnet's physical ability as an electromagnet. The transformer energy of the electromagnet is primarily magnetic, whilst the primary energy of an electrical cable is electrical. Magnetic energy clearly has a greater physical capability compared with electrical energy within the cable. The cable energy lacks mechanical or physical capabilities.

The vibrating hum of a transformer is generated by the mechanical wave of the frequency which is generating a physical wave within the magnetic flux particles of energy (i.e. the physical vibration of the laminations and other metallic parts of the transformer). Magnetic energy has physical properties which are 137 times more powerful than gravity, and later, I'll give you further evidence of the existence of the mechanical wave of our Earth which pulses out a physical frequency wave from the core at 8–12 Hz as recorded by science.

Again, just like the violin, if you pay close attention when listening or feeling the physical vibration and sound of any transformer, you will notice that there are three different amplitudes of the hum generated by the frequency wave of 50 Hz in the transformer. *I believe that this may be why you hear the hum of the transformer going up and down.*

Resonance in the Human Body: The Butterfly Vibration

It's a well-known fact that each cell of the human body is made up of a number of atoms which hold an electrical charge in much the same way as a battery cell and that these cells of the body can become depleted of that charge, thus suffering chronic fatigue. In much the same way, a car battery can become depleted of its energy. Electrons possess a negative charge, protons have a positive charge, and neutrons are said to have no charge. It is that charge on the atomic particles listed directly above that become depleted, causing the body cells to also become discharged.

A number of atoms make up a molecule, which will exist in (more or less) an accumulative or combined frequency (a little like a number of musical tunes that play together to make up a melody). A number of molecules make up a human cell, which are identified as blood cells, marrow cells, bone cells, skin cells, brain cells, etc. The cells become weak in their charge, causing chronic fatigue, but I can tell you that the marrow contains the most important nutrients, plus the actual bone itself.

In the class of naturopathy, we were taught the science of the body, where the human body emits a frequency which can vary anywhere between 4 Hz up to 1,600 Hz, which is a huge variation. One of the ways that we can change the frequency at which our body vibrates is by what our body digests. Two very good examples can be experienced in

- sugar, which will raise the body frequency or vibration, and
- herbs, which will slow down the body frequency/vibration. (Try sprinkling some mixed herbs in a glass of water and then drinking it. Try then to stay awake.)

We were all taught in naturopathy that the word 'emotion' is derived from two words, which are 'energy' and 'motion'. I have since learnt that every frequency at which the body vibrates has a direct reflection or effect on an emotion of our mind. I also have the opinion that when a monk strikes a gong whilst they meditate, which is in tune with the 'frequency of peace', they're constantly trying to align with this frequency so as to attune to the perfect or resonant frequency for their body and consequentially their mind. I'm sure that you will agree that the energy and thus emotion of a monk has a somewhat slower physical pace and is a peaceful energy flowing to the body and mind.

Our solar system also generates frequencies which project onto our Earth and affect the majority of us emotionally and thus socially. The greatest example of a frequency being produced by our solar system and thrust on our Earth can be seen by the emotion of the general population during the 1960s. I won't put any preconceived ideas into your mind, but ask yourself now, 'What was the most obvious emotion of the greater percentage of the community during the 1960s?'

The 1960s were forty years before the planet alignment, and I have a very important theory to tell you about 2040, which is forty years after the alignment. *It will be a mirrored or reversed planetary period to the 1960s. The emotion generated in 2040 will be the opposite of 1960, and I'll describe it in the later chapter on social effects of these lines.* I'm sure that you will agree that, ever since 2000, the rate of violent crime has been ever increasing, and this is something which will worsen all the way up to 2040, but I will have more soon on that later to help you understand how it occurs.

During my naturopathic observations over the last thirty years, I have theorised that when in conversation, the human body can vibrate or tingle **whilst talking,** and this is the result of the two bodies tuning in, thus becoming in resonance. It can occur (but not every time) in just a limb or just one area of the body or the entire body. The result is an exchange of energy, and I believe this is just one of the many causes of chronic fatigue when you unknowingly give away your energy. *I refer to this resonance occurring in each body as the **butterfly vibration**.*

Where and when the 'butterfly vibration' occurs, it is just that part of the body (or in some cases the entire body) which is vibrating due to being in perfect resonance with the other person. The transfer of energy during the butterfly vibration is just one of another causes for the cells to become discharged and lose their energy, but it is actually a new body science that only I have described and introduced here today.

I need to add one last fact in regard to the butterfly vibration. During a number of hours over a couple of days, I had also experienced the butterfly vibration within the area of my heart chakra and my base chakra. *At first, it felt very mysterious, until I remembered the teachings about the chakra system.* The most common chakra system exists in **seven** different frequencies which displays seven different chakra colours, and this memory came as a relief during those peculiar experiences. I'm sure that I'm not the only one to have ever felt the butterfly vibration in any of the seven chakras, but as I said, I have also felt a mild butterfly vibration in my heart chakra on just a few occasions. The cause of the butterfly vibration for me during those individual hours and particular days is due to frequencies being thrust on our Earth by the solar system and is described with some detail in the later chapter on social effects by the lines.

My experiences with the base chakra vibration on those particular days were relative to the butterfly vibration but must actually be specifically identified as a base chakra vibration by the red chakra. One of the dates and times of the base vibration which occurred to me was at the very moment that they announced on the radio the passing of singer Michael Hutchence. God rest his soul. By this statement, I am implying that it may strongly affect some of us a great deal more in the mind. I need to also say that during the weeks before experiencing that base vibration, I was also suffering from an incredibly deep, depressed emotional state, but today I am the happiest person on the planet.

In a later chapter, I further describe (in great detail) how emotions are generated and felt when particular planets are in particular positions, plus the periods of the year when that energy is most felt as it is thrust on our Earth. Just briefly now, I must inform you that Jupiter is said to be 395 times bigger than our Earth, and Saturn is recorded to be 95 times bigger than our Earth. Saturn is a big gaseous ball of rubbish and therefore absorbs energy away from our planet.

Someone once said to me that the term 'satin' in the Bible was derived from the planet Saturn. I don't know how true it is. I'm not into theology because I cannot read the Bible and have no desire to try. I am told it is a very violent book.

**Every frequency of the body is directly related to an emotion of the mind.** If you don't believe that sugar will change your body's frequency, then try some chocolate late at night when you're already half asleep, some alcohol in the morning after you wake up, some of those herbs I mentioned above, or some marihuana that the government now grow, sell, and tax to cure illness. I also have no attraction to that stuff either.

Now that we are understanding frequency and resonance a little better, let's now look back to our sun's creation. Remember that hydrogen gas is only a single atom which exists or possesses just one frequency and is not a combination or a number of frequencies, like that of a molecule of gas. The actual junction within the crossover of these lines very obviously generates the identical frequency of the atoms which it attracts. The junction is attracting single atoms of hydrogen, plus a very small percentage of some other gases due to magnetic resonance. The frequency of the junction is in tune with the gases which it attracts, but as a molecule, there is a very high percentage of hydrogen gas, around 97–98%. To understand this molecule of gas (in lay terms) a little better, just think of a band where all of the members each play their notes to make a melody.

These are very basic facts about resonance and hydrogen atoms of gas which are **very** important. Every one of the 400 billion sun-stars in our galaxy possesses the identical magnetic resonant frequency to that hydrogen-based molecule, having been formed under the exact same mechanism, and I'll show you how that occurred.

Stage 7: The Magnetic Pump

When I was called to work in the engineer's railway yard as an electrician, that first day at lunch, I saw the old book titled *Science Dictionary*, and I picked it up out of sheer curiosity and began referencing the word 'magnetic' when I stumbled on 'magnetic pump' with the following description: 'When magnetic lines cross over, they generate a magnetic pump, begin attracting atoms of gas, and heat them to 1,000,000 deg.' It didn't actually specify which atoms but merely quoted 'atoms of gas'.

At home that afternoon, I looked into the internet for our sun's temperature and found that, within the centre of the sphere, the reactivity reached estimated temperatures of 1,000,000°C, but the surface is only 6,000°C due to the extremely cold temperature which invokes a restricted burn. The area of space surrounding our Earth today is recorded at super-sub-zero temperatures of around **-300°C,** but this area is now **heated space** generated by our sun, which prompted me to ask, 'What would the temperature have been before the existence of our sun?'

The differences in temperature between our sun's core and the surface today are the result of the surface experiencing a controlled burn which is restricted by the super-sub-zero temperatures of space that the surface must first overcome before the possibility of 6,000°C being achieved. Science, for decades, have explained that every sun in exactly the same formation and these two temperatures from both the old science dictionary as the magnetic pump at 1,000,000 degrees and the core temperature at 1,000,000 degrees seem impossible to be just coincidence. Today there appears to be two different opinions from early science and from today's science.

Do you think that the conclusion of the two temperatures above very obviously was the result of the same laboratory results recorded by early science? Further, could it be that the actual temperature of our unheated space during the period when our sun did not exist must have been incredibly low or incomprehensibly low? However, taken from the difference in temperatures that the surface (6,000°C) must overcome and the recorded core temperature of 1,000,000°C, it may be possible to estimate an unheated space at around -10,000,000°C, but it's unlikely.

The true temperature of our unheated solar system will probably remain a mystery, but it will have been very good to measure the temperature around Pluto near the boundaries of our solar system when NASA recently flew the *New Horizons* spacecraft there in 2015 after leaving our Earth in 2006. To possibly gauge the temperature of an unheated area of space, I've contemplated an estimated area around Pluto with a conservative temperature of -200,000°C or greater.

Just so that you realise the potential of super-sub-zero temperatures, let me tell you just one small fact I recently learned. I read a report by science describing **cold fusion** which occurs between two metals in super-sub-zero temperatures by simply touching. It seems unbelievable but true.

Some people I spoke to suggested that my description of a magnetic pump is clearly nuclear fusion that occurs in nuclear reactors. I have to admit I'm not actually a learned person in regard to science. This form of magnetics seems to be my forte, having been taught it and worked with it as an electrician, although I do have a number of other post-trade electrical qualifications in electronics; computer logic controllers, which are computers used to run buildings and factories; electrical contracting; and a few other electrical ancillaries. Besides my extensive electrical trade competencies, I also studied and gained an open-class gas fitting license, air-conditioning gas license, telecommunications licence, plumbing license, and pest control license along with a great number of other post-trade studies, including every welding technique and metallurgy. I achieved a great number of competencies in a varied number of trades because I wanted to do pre-purchase reports in the housing industry. Today all these competencies are applied here in the book. I learnt a great deal about "frequencies" in my telecommunications studies, and what I learnt in pest control was also very important here in the book as they applied in a later chapter about evolution of the species. More about me later.

Stage 8: The Vacuum in Space

Science records the area of space around our Earth in a vacuum but also with a very high percentage of hydrogen gas, plus a small percentage of helium, along with a very low percentage of various other gases, including oxygen and carbon, which together make up the molecule I spoke about. If every sun is attracting, heating, and burning an incredibly huge amount of hydrogen and other gas atoms at a rate of almost instant combustion, then the supply and demand on these atoms must be almost infinite or nearly impossible to keep up with for 400 billion sun-stars in our galaxy, the Milky Way. The source of the atoms is very obviously an incredibly long distance across our galaxy or maybe beyond our galaxy up to a gazillion light years from the suns in faraway, neighbouring galaxies.

I believe the area of space around our Earth is under a vacuum due to a number of different causes. The following lists the side effects of combustion:

- The incredibly high demand on gas atoms where the atoms possess mass, thus slowing down their travel through space due to resistive values on the front face of the atoms
- The almost infinite distance each atom must travel to eventually be instantly combusted
- The loss of their mass within the area of space around our Earth, after the atoms of gas have been instantly combusted, taking place at 400 billion sun-stars, which I believe is contributing to a great deal of the vacuum (i.e. after an atom has been combusted, there is not enough time to replace its mass)

The vacuum that exists in space would very likely be a great deal less if there weren't so many atoms being combusted after the mass of each atom is lost. *I'm saying that atoms possess a large enough mass that the loss of mass and the speed of replacement generate the vacuum.*

The description in the old science dictionary of a **magnetic pump** wasn't specific enough, but also (looking back today), I was extremely dyslexic that day in the lunchroom. You see, the effects of my dyslexia seem to change nearly every day or, in fact, hourly—that is, until lately as my dyslexia has improved whilst writing and reading this book called *The Holy Grail of Science* during the last decade. It has slowly improved over the decade, now being able to read better than that day with the old dictionary. On the day that I read the old science dictionary, it just didn't click anywhere near like the lightning bolt that struck me on the day of discovering the master key. That split second was a moment that I would certainly never forget.

Upon realising that the **master key** had been just two lines crossing over, I then began wondering where the two lines had actually come from. My next conundrum about the lines had begun AGAIN. I asked myself, 'How had the master key (as the crossover of lines) come to function as a magnetic pump?' Well, the answer is actually extremely simple. Let me to show you now. Grab your hat again, and let's go

Stage 9: A Crossover of Just Two Lines and Where the Lines Had Come From

My contemplation began at home. I sat there just staring at <u>the new master key</u> of two crossed-over lines and no **atoms**. I was just staring again at the new diagram. ***Just two lines?*** *But where had they come from?* It began to eat away at my mind.

I was now beginning to totally lose my mind. At first, it was just hour after hour of being obsessed with two lines, and it seemed to go on for days, lost in contemplation and utter confusion. I could actually feel the answer was close. I could feel that I just had to figure out how the lines had come to be crossed over, and so I continually asked myself the question over and again. 'Where had the lines come from?' I'd almost given up when . . . well, actually, I had to give up.

It was Friday, a full week of contemplation about the lines and also my house renovation, plus paid work commitments. The week had come to an end, and it was now time to pick up my son for our weekend bonding. The answer to the question was delivered that Friday afternoon, but just like the old science dictionary, this answer had also failed to sink in. It had to do with atoms of gas.

This answer was literally just around the corner. My 7-year-old son, Ben, sat, waiting to be picked up from school on my way home from work. He was about to bring it home. As I said, the answer was literally just around the corner and was delivered to me through my son's attentive mind the very first moment he arrived home, but I missed it—well, at first.

It all began when he saw the white board with the master key diagram on it at home, and he asked, 'Dad, what's that drawing?'

I replied, 'Oh, just a new theory that your dad has about our sun.' I was going to leave it there when he suddenly became very excited and began telling me about his lesson on our sun that very same day at school. It was all about our galaxy and the 400 billion other sun-stars that inhabit it.

It was his next statement that gave me the answer, but I mustn't have heard it, and just like the old science dictionary, the answer had just slipped past me. His reply too had gone straight over the top of my head. It must have. This was Ben's master key, said in just few words, which I should have listened to. 'Dad, just next door to us, there is another sun-star which is 100 times bigger.'

In that exact moment, I was initially overwhelmed but not about the fact he had just told me the answer. I was amazed that he was learning all of it in grade 6, and so I simply replied, 'Oh yeah?'

He continued to tell me what he had learnt in school that day. 'Then next door to that, there is another sun-star that is 1,000 times bigger.' Ben was very excited whilst he told me the great news about our galaxy, but I was actually dumbfounded that he was learning all this in grade 6. Grade 6, folks!

The names of these two neighbouring suns really are irrelevant, and so we won't worry about them. I left it at that. We both enjoyed our weekend of special time, and life continued, UNTIL THAT MONDAY WHEN HE RETURNED TO SCHOOL. Like I said, the answer was delivered to me, but I obviously didn't hear or see it. Can you see it now?

Some people will try to convince you that lightning never strikes twice in the same place, but we all now know today that this is an absolute load of rubbish, right? Lightning can strike twice in the same place or maybe up to several times. Remember the American forest ranger who was struck seven times and eventually committed suicide? God forgive his soul.

*Please also allow me now to take the opportunity to make another personal statement right here. If you too are separated and have one or a few children, you truly need to cherish every moment you have with them. They grow so fast, but more importantly, you must also help **them** get through separation anxiety by contemplating what they go through. That's all I wanted to say. Thank you.*

Let's get back, shall we? It's beginning to heat up with all this talk about the sun.

On that Monday, Ben returned to school just around the corner after another sensational weekend of fun when I literally sprinted my way back home. Another lightning bolt had struck, and I was feeling it. I had to get home to the new diagram of the master key with just two lines which crossed over.

I was standing at the white board, adding my vision to the master key diagram. In fact, I was drawing those two neighbouring suns that were 100 and 1,000 times bigger than ours that Ben had told me about. I was drawing them into the diagram like some sort of crazed animal when I saw the extremely logical simplicity of two neighbouring suns. It took me about ten seconds to discover where the Lines had come from.

This next discovery was huge, one colossal step for man but one giant leap for mankind. I'll describe it now with immense gratitude to the dedicated teacher who taught all this to my son and, of course, to that great young man and son of mine for paying attention in class. So have you still got your hat on? We're all about to discover how our sun-star was formed. HERE WE GO.

You'll never believe how truly simple it all is, but I'm about to show you how our sun-star was formed. If you don't believe it, then close the book and go back to your life. Let's go. You're about to experience a global phenomenon in your life that will change the way all you think about science.

Stage 10: The Moment Our Sun Was Created

Our sun and each of those two neighbouring suns, plus every other sun in our galaxy, exist in accordance to the diagram. Each is a magnetic pump. The system of a magnetic pump is repeated 400 billion times over with rotating magnetic lines radiating from a reactive centre sphere or ball. The key here is 'the rotating lines' or 'arms' that reach about fifteen billion kilometres across our solar system today, but the thought beckons. How far did or do the arms of the neighbouring solar system reach?

All 400 billion of them are so incredibly synchronised in cycles. It is the synchronisation of these arms or neighbouring lines which is important. It is the synchronisation which has to be seen in the following chapter.

The neighbouring suns that are 100 and 1,000 times bigger logically have huge—no, HUMONGOUS lines in comparison to our sun's solar system, which may, in fact, span further into regions of space, but will it be possible for any sun lines to go beyond their boundaries into the boundaries of other systems? An extremely important factor to keep firmly in mind about every sun and solar system is the **synchronicity** which can be seen in the following reveal about our galaxy, the Milky Way.

The Timely Synchronisation Occurring in Our Galaxy to Every Solar System

The synchronicity of our sun's lines and every other sun's lines is inconceivable, and you will clearly see this from the following. It begins with the fact that our sun (as the reactive sphere) and our solar system's boundaries (which are defined by the energy of the four lines) altogether rotate once every four years. The fact that our solar system takes four years to rotate once is governed by the number of poles our sun possesses, which is the very same rules of an electric motor that govern the RPM.

Every solar system across our galaxy exists in the same configuration of four lines, and they all therefore rotate at the same frequency. Four years can be seen as the frequency of each solar system, especially the two neighbouring solar systems. I'll show you the evidence of a four-year frequency later. The four-year frequency period evenly equates into 500 years, which is the period that the giant science atlas quoted for a planet alignment. **The orbital frequency of each planet in our solar system also equates evenly to 500 years**.

These are the facts that make it possible for the planet alignment to occur, and the orbit frequency of each planet is listed below.

Mercury - 87.97 days
Venus - 224.70 days
Earth - 365.26 days
Mars - 686.98 days
Jupiter - 4,332.82 days
Saturn - 10,755.15 days
Uranus - 30,687.15 days
Neptune - 60,190.03 days

Pluto is recorded to take 248 years to orbit our sun, which may be slightly inaccurate in accordance to the laws of planet alignment.

Every 500 years, our entire solar system undergoes a total planetary alignment where every planet for just a split second forms one single ascending line. I must tell you that it is not just a single line of planets. It is the magnetic energy of every planet which aligns. The planets also become slightly staggered due to the amount of time they have all been orbiting. It is the magnetic energy of each planet which forms a line. Incredibly but very truly, the alignment takes place for just a split second in a single moment of our solar system's synchronicity every 500 years, but there's a great deal more about synchronicity; in fact, there are incredibly synchronised events also seen in the following.

The alignment now during our time is recorded by science to form one single line of planets running out into the northern hemisphere of **our** solar system, and the next planet alignment 500 years from now (AD 2,500) will take place in the next clockwise quadrant, aligning in the eastern hemisphere of our solar system with more synchronisation. It's a little difficult for me to describe, but imagine that our entire solar system also rolls around the inner wall of a balloon, and all our planets keep forming alignments into next quadrants running out towards the inner wall.

I have concluded that, whilst all our planets are forming each line in each next quadrant, here in our solar system, at the same time, one neighbouring solar system is aligning all its planets with ours in their quadrant but only every 2,000 years. (I'll show you the evidence of that later.) When all the planets in our solar system form this line, it is along the energy of one of those four lines of the diagram. The energy we all experience every 500 years is an amplified energy, but the energy generated now by the doubled-up energy (including the neighbouring aligning system) is extremely strong every 2,000 years. I'll describe some of the prima facie evidence in some detail in a later chapter about the evolution of the species.

Some neighbouring solar systems are extremely large (in accordance to my son's grade 6 lessons), and some are absolutely humongous, like the neighbouring sun-star which is 1,000 times bigger and its solar system. The facts described in the example of the balloon clearly explain the colossal amount of synchronicity for every solar system as if the entire galaxy were some kind of huge clock. And well, I'm here to tell you that it actually is a humongous solar system clock, but the facts don't stop there.

Our solar system's clockwork is slightly ambiguous (at least for me) in regard to our leap year as well as Pluto's two-year conundrum. I believe that the extra day every leap year, which is **every four years,** was a consideration by early science when contemplating our Earth in the next quadrant and a further quarter of a turn into the next alignment. Each alignment reoccur in the next quadrant in a clockwise direction and so on every 500 years. I'd just like to say that if science could read the exact dates that the lines of the diagram are passing by our Earth, we may have discovered a better way to read the exact time of the year or exactly how long a year is.

With regard to every planet's orbital frequency, I sometimes wonder whether each planet's pulse, which has been measured by science, will be a strong factor in also governing each planet's orbital period, but those facts will also be expanded on in a later chapter.

There are two extremely important facts about the 500-year alignment cycle. The first—which I'm sure you can clearly already see—is that, during the 500-years orbit, one or two or maybe three or so planets may end up orbiting in a slightly unsynchronised movement for just a split second. *During that final split second of planet alignment, an adjustment occurs <u>caused by the strong 'aligning magnetic energy' as it pulls every planet in sync by the strong magnetic force</u>.* This aligning adjustment suddenly occurs to some of the smaller planets as they ALL pass one another with momentum.

I refer to this realignment adjustment mechanism of final synchronicity when the magnetic alignment energy is placed on every planet as the **synchronising planetary mechanism**. *Besides the synchronised planetary mechanism, there is a secondary mechanism occurring during an alignment which I refer to as the 'magnetic planet **re-energising** mechanism'.* Any planet which may have weakened in magnetic energy whilst orbiting during the 500 years will be re-energised by the stronger energy of an extremely stronger or larger planet whilst passing each other during the alignment. This is due to inducement proximity magnetism. This re-energising by inducement is by a stronger planet into the weaker (see facts about inducement above). The re-energising mechanism has a very definite effect in recourse on each planet after it has been re-energised. Clear evidence of the planets being re-energised is described in a later chapter about extreme weather conditions in events that occur AFTER an alignment throughout our Earth.

Our moon's cycle is the most noted cycle of synchronicity, which is every twenty-eight days. This has occurred over and over again in a synchronised pattern of events for millions or maybe a billion years, which also has a secondary effect on our oceans during the tide cycles. Earth's night and day cycle of twenty-four hours is a smaller cycle. All the extremely timed events mentioned above are natural cycles, including each 500-year alignment cycle. The 500-year planetary cycle forming in each next quadrant, the yearly orbital cycle, the quarterly cycle of the four seasons, the 28-day moon cycle, the quarterly cycle of king tides and each of the neap tide pairs, the daily tide cycle, and the 24-hour day/night cycle press home the strong fact of **synchronisation** occurring everywhere, especially to the two neighbouring suns' lines. Remember that each neighbouring system takes four years to make one single rotation. This means that, every year, the neighbouring lines will cross over, which is a quarter turn for a neighbouring solar system. *So the next time someone says to you that 'there is no time that exists in space', suggest to them to read The Holy Grail of Science.*

The list above does not include any of the cycles influenced by the diagram lines, which are the flowering of plants, the coral reef spawning cycle, and other cycles of Mother Nature. The cycles of cyclones, hurricanes, typhoons, and tornadoes should also be included into the cycles caused by the lines. I tried to only include astral and planetary cycles and have not included other man-made cycles of time such as millennium, century, decade, month, week, hours, minutes, seconds, and nanoseconds.

All this helps us realise just how incredibly colossal and super-synchronised our galaxy and all its solar systems really are. It is truly incredible that this synchronicity very much includes the neighbouring suns' lines and neighbouring solar systems.

Decades before and after each 500-year alignment cycle, all our planets begin to form groups. At first, these groups appear quite loose, but these 'loose groupings' begin many years before the alignment. After the planets orbit past one another, these loose groups continue on their orbit; and in the next stage, those planets are in the same quadrant of our solar system where they group again but in a tighter formation. Each time a grouping occurs, it's governed by the orbit period of the planets as they approach the 500-year alignment. Eventually, the planets form a tighter group before the actual alignment; and each time the group is formed, the energy becomes greater, depending on the tightness of the formation of each group, very much like a bunch of magnets that are positioned close to one another. After the planet alignment, these identical groupings occur again, but the planets will now be in mirrored positions in the same amount of years but with one huge difference, which I will tell you about soon.

During these groupings, the energy on our planet Earth increases. Jupiter is around 395 times larger than our Earth, which means that it projects a lot of energy, and Saturn is also around 95 times bigger than our Earth. The intensity of the energy during these groupings really is governed by Jupiter and Saturn. Jupiter takes around 11.5 years to orbit, and Saturn takes around 30.0 years. When these two planets are both near our Earth, it generates a lot of energy and has the potential to occur every 30 years when Jupiter is nearby, but the energy also increases when the other planets form close groups. As all the planets approach the actual alignment, these planets become even closer or form tighter groupings.

Previously, I briefly mentioned that the weather conditions on our Earth intensify after an alignment. You may have already noticed that the weather patterns have intensified ever since the year 2000. This is due to the 'magnetic planetary recharge mechanism' that occurs during an alignment when it recharges each planet, placing each planet in a stronger position or stronger overall state of energy, in turn generating a stronger weather effect. This recharging occurs especially when the planets align, but bear in mind that the planets are aligning along one of the diagram lines, and the line also has a tendency to increase the energy. The exact details of the change in weather before and after an alignment, plus how it is generated, is described in detail in a later chapter.

There is also what I identify as 'the normal energy period', which is 250 years before and after an alignment when every planet is scattered, but the overall energy begins to increase about 120 to 150 years before the alignment and stays at this level for 120 to 150 years after an alignment. The level of magnetic energy generated during the actual alignment is very intense, and the intensity of the energy is directly reflected in our communities. Quite simply, the planets can be viewed as simple magnets during an alignment and will generate a stronger energy whilst grouped together, thus amplifying the total energy.

I believe that the alignment which occurred during the turn of this millennium had a strong, aggressive social effect, resulting in increased violence around the period, which was obvious in the terrorism during 9/11. God rest those souls. I can only assume that the alignment and intense energy occurred months before 9/11 when the people involved would have had their terrorist meetings or communications about the event. It's important to understand that increased or intense energy can show itself in the community and can be reflected as stress, increased business transactions, inventions, or anger. Quite simply, the high energy during the period leading up to an alignment, during an alignment, and after an alignment also causes side effects in the weather on our Earth and the community as a result of the increased magnetic energy. (More on that later.)

I contemplated synchronicity a little further to arrive at the following description of exactly how our sun was created. The area of space where our sun exists today was once empty, and I began to contemplate that maybe those humongous neighbouring sun lines projected out beyond their boundaries. In fact, they did and still do today. They projected beyond their boundaries when the area of space where our sun and solar system now exists was empty, and the neighbouring lines entered into that empty area and then crossed over, creating a magnetic pump. This occurred over and over again every year, which was every quarter turn for both of these neighbouring sun lines that were synchronised to cross over.

Remember that our sun and the neighbouring suns' lines take four years to complete one rotation, and the synchronicity of the lines still occur. Therefore, a quarter turn every year is when the lines meet again during synchronisation. Then eventually, they permanently formed a magnetic pump with an incredibly reactive centre point which began attracting gas atoms from the very first moment they crossed over, eventually heating them to one million degrees, finally forming a permanent magnetic pump.

*I felt that this mechanism of lines crossing over needed to be identified better, and I named the crossover after my great-grandfather (who has helped me a great deal in my book and my life) as **the sun-star crossover**.*

His name was Simon Cross, who performed as a stage clairvoyant amongst other talents under the name of Professor Nomis. Nomis is his name Simon written backwards. He performed for Queen Victoria as the support act for Buffalo Bill's Wild West show when they travelled the UK, and he was a one-man show that mostly carried out a ventriloquism act. A little more detail about the sun-star crossover is listed below.

When those neighbouring suns were young lines, their lines were obviously straight. The fact that they were shorter meant the reactive centre of each sun would have had to increase in reactivity first before the size of its lines would have been able to grow, including the length of the lines, whilst still rotating. I actually estimate that the growth of each line had to continue for many billions of years before the lines of each neighbouring sun were eventually long enough to be able to cross over, which they did. All the while, both of the neighbouring suns' reactive centres would have also been growing in size and energy, with their lines becoming not just longer but also magnetically stronger whilst the energy radiated out along them.

The first example above of resonance was occurring to water waves, the second was wind waves, the third was sound waves, the fourth was electromagnetic energy (primarily as magnetic) waves, and the fifth (my new science for resonance) occurs to the human body, which is also actually just another example of electromagnetic waves. **Now allow me to show you one last theory that I have on resonance as the compressed energy occurring with a magnifying glass.** This isn't an overly clear example of resonance, but it does describe the compression of photon light particles of energy as they are projected out of a magnifying glass. To best understand this next example of resonance, it might be best to elaborate on the element of light particles.

When I was at a recent concert, I was almost hypnotised by a laser beam in the dark auditorium.

When a beam of light travels through a dark concert hall, it is doing exactly that. The beam or laser of light is travelling out of the light fitting and flowing through the air until it hits the back wall at the speed of light. Let me describe that in greater detail.

When the laser is initially turned on, the light fitting supplies the source of light in a beam or flow of light particles to the back wall. If we were able to video the initial switching on of the laser beam which is travelling at the speed of light and slow it down, we would see the front face of the light beam travelling from the front of the light fitting all the way to the back wall of the concert hall or amphitheatre. The actual stream of light particles doesn't stop there. There is a constant flow of light particles from the front of the light fitting to the back wall. If we were able to slow down the video of the light beam, we would see a constant stream of light particles flowing from the light fitting along the beam of light (in the dark) until each particle reached the back wall but travelling at the speed of light (of course).

Little do we realise that a beam of laser light doesn't travel any faster than a general light beam or a beam of light coming from a normal stage light. The laser light beam particles are, however, more concentrated. Plus, a normal light in the middle of the room (as in the light above you in your lounge) has light particles flowing from all around the source of the flow outwards to every part of the room and is not so concentrated.

This may help you better understand the next theory on the light beam of a magnifying glass.

Just quickly and quite simply, when sunlight is projected out of a magnifying glass, it is (more or less) 'compressed into a fine point' when the light particles are channelled into a V of sunlight energy.

During my last edit today, just before publishing, I am beginning to strongly consider that a magnetic pump is generated by a magnifying glass where the crossover of sunlight particles occur within the V of the beam just outside the lens of the magnifying glass. This is the component of magnetic energy in lines within the light energy crossover.

In the following chapter, I very simply describe a more detailed example of magnetic resonance energy, which is the reactive mechanism within a magnetic pump. PLEASE REMEMBER THESE EXAMPLES OF RESONANCE. As I said, I will later describe further evidence of magnetic resonance, but we're taking this slowly.

This mechanism of neighbouring sun lines entering and then crossing over did create our sun. The mechanism of neighbouring suns' magnetic lines entering into our area of once empty space was exactly how our sun-star was generated but was also how our two neighbouring suns were created. This mechanism of neighbouring sun lines projecting out and crossing over occurred in a domino effect that created sun after sun after sun after sun after sun, in fact all 400 billion of them and still growing. This domino mechanism of sun creations also helps us realise just how infinite and especially how precisely timed the overall function of synchronicity really is throughout our galaxy.

As our galaxy grew over time, the crossover of sun lines to create new suns (as magnetic pumps) by every other neighbouring sun occurred in a period when they would have had to have straight lines. Some of those 400 billion began curving comparatively early after their creation during the total process. What I am gently trying to say is that this mechanism of neighbouring lines crossing over into further neighbouring and empty areas of space is how our entire galaxy, the Milky Way, was created. Plus, depending on the configuration of the neighbouring galaxies, it may also be how THE ENTIRE UNIVERSE WAS CREATED.

I hear you ask now, 'Where or when or how did the first crossover in our galaxy ever happen?' I actually theorise that maybe the first crossover of lines came from neighbouring galaxies, but it's just a passing thought I have right now, and I don't really know for certain (at the moment). Our universe really is infinite.

I was describing the crossover recently to a man in real estate who had a science degree, plus the details of how I had theorised every sun's creation and other facts surrounding my theory. He said, **'Einstein was reported to have looked to the heavens to proclaim his mind to God when he said, "I want to know which came first, the energy or the atoms."'** And I'll answer that conundrum soon.

So our sun was created after neighbouring lines crossed over and formed a magnetic pump from just two neighbouring lines, which eventually reacted enough together. The reactivity then formed as the reactive sphere of our sun and then began functioning as four lines.

The Rotation of a Wire Clothes Hanger Straightened and Suspended from the Ceiling

One day I was experimenting with a wire clothes hanger when I straightened it out and suspended it from the ceiling on a single thread of cotton. I wanted to make a compass so that I might be able to see it's reaction every three months when the lines of the diagram floated through our Earth but I didn't have a magnet in my home that I could stroke the hanger with so as to induce it with magnetism. I had not ever charged the hanger with magnetic energy from a magnet by stroking it, which would have definitely turned it into a compass needle, but had my fingers crossed that it would react slowly like a compass, and it did. But it did a great deal more than that.

Initially, it almost immediately moved into a north–south orientation, but it didn't stop there. Majority of the time, the hanger would point north–south; but a lot of other times, it would rotate from left of true north back through true north and then to the right of true north and back again. On the days that the king tide energy was present, the wire hanger would rotate at a greater rate of knots during the peak of the king tide. On king tide and neap tide days, it pointed almost all the way to the west and back again through north, almost all the way to the east position, and then settled again at magnetic north. It did this over and over again throughout the day of the strongest tides. What we all actually need is a video recording of the hanger over a long period with a timestamp in the video so as to match the rotations up against the tidal chart because the hanger appeared to make significant rotations during general tide hours plus stronger rotations during the period before the king tides and much stronger rotations every three months when the king tide was in play.

As previously mentioned, every sun's lines and the reactive ball take four years to make one full rotation. Those neighbouring sun lines still project here today into our solar system and our Earth. The energy of the neighbouring lines still enter into our solar system as a single line every twelve months when one of the neighbouring lines will point into our solar system. Rotation of the hanger is yet another reflection of that energy on those days by 'significant and strong rotations'.

I also noticed by coincidence that on the days when significant and strong rotations occurred, there were reports of tornadoes or earthquakes on our Earth. The hanger may, one day, allow us to be able to predict tornadoes or earthquakes; and for this reason, I named the hanger as an instrument of measurement called the Cox meter.

The following describes where the magnetic energy that rotates the hanger comes from. This magnetic energy comes from a great number of sources, mainly our moon's radiating lines which project onto our Earth and those from <u>our</u> solar system's planets, our sun's four radiating lines, both of our neighbouring suns' four magnetic lines, and the open- and closed-circuit lines within our solar system that our Earth orbits through.

The Cox meter or straightened clothes hanger very obviously reacts as an indicator to high energy lines being developed during their projection onto our Earth. Maybe in the not-so-distant future, the Cox meter will become some kind of quake meter or an energy indicator of some design, working together with seismographs. I can also tell you now that, due to all the lines being so extremely synchronised, they are therefore mathematical in their projections. All quakes also actually exist in a mathematical pattern that can be calculated and predicted. In the past, I have used a calculator and calendar to make accurate predictions of quakes, and it can be done again with the aid of others who understand these simple principles. I'll fill you in later on the exact details, but I can also tell you that quakes are a result of our sun's lines, plus neighbouring suns' lines, which meet head-on during projected synchronicity.

Our sun's lines were generated during a crossover of two synchronised neighbouring suns' lines; therefore, our sun's frequency is also identical to each neighbouring sun's frequency. Our frequency and the neighbouring sun's frequency are also identical because the rotation frequency of each sun is dependent on the number of lines or poles which governs it, just like an electric motor. Because of the synchronicity of two neighbouring lines which generated our sun and its lines (in an exacting moment), all these sun lines are now synchronised, and so it is therefore not just the frequency which is identical but also the synchronicity or timing of the lines.

Recently, in 2018, there was an announcement of a huge planet (1,000 times bigger than Earth) observed just at the outer region of our solar system. When we consider the sizes of neighbouring suns and their lines (100 and 1,000 bigger), the size of this neighbouring planet suddenly appears very logical. Yes, there are neighbouring solar systems which contain planets, complete with moons, and the discovery of that huge planet (at the outer edge of our solar system) is clear evidence of this.

The following is a description of how all our suns were created in the Milky Way. I mentioned that the neighbouring suns' lines would have to grow in length before they could finally cross over. Whilst the reactivity was increasing, the lines of the neighbouring suns grew, and this growth was occurring to every sun. Eventually, after the length of the two neighbouring suns' lines grew long enough, it was then possible to create the next sun, which was ours. This occurred to sun after sun after sun after sun under the mechanism that I refer to as 'the domino effect'.

So now you can see, according to *The Holy Grail of Science*, how science has said for decades that our galaxy is infinite due to it forever expanding. Space exists in an extremely high percentage of hydrogen atoms due to the resonant frequency of each of the 400 billion sun-stars which are attracting, heating, and burning megatonnes of hydrogen gas atoms almost all at once.

The domino effect, which is the **crossover of lines,** occurred first, followed by the attraction of the resonant hydrogen atoms (plus the other gases). **Don't you think that these facts answer Einstein's conundrum? Energy was first and then the atoms.**

The magnetic pump functions in the following manner. Remember that the lines are composed of electromagnetic energy and that they are primarily magnetic energy, but at the same time, electrical energy is also present. The lines very obviously do have a small percentage of electrical energy despite being primarily magnetic energy. It's that small percentage of electrical component that holds the key to magnetic resonance.

Of course, the frequency of any energy which is primarily magnetic is directly dependent on the frequency of the minimal supply of electrical energy which is also present. I estimate the frequency of all magnetic energy as very low. The frequency pulse which our Earth emits from the molten magnetic centre core would be a strong indication of what the average frequency of magnetic energy could be. Therefore, the mechanism of magnetic resonance occurring within the magnetic pump has a very low pulse and is best described below.

Again, let's take our mind back to that first time when we looked at a magnet being placed under a piece of paper with iron filings sprinkled on top. Remember that the lines of magnetic energy were reflected in the magnetic lines of iron filings, and this is the simplest way to understand that there are absolutely mechanical and physical properties to magnetic lines. Also, remember that magnetic energy is recorded to be 137 times stronger than gravity, and I must add that gravity is **_not_** described as having a pulse, but that doesn't mean it doesn't have a pulse, right? The pulse that has been measured whilst being emitted from the core of our Earth could, in fact, be the pulse which is emitted by gravity.

The molten magma core would be the major magnetic field which would be surrounded by closed-circuit lines that would begin somewhat close to the centre core but would also spread throughout the middle core, plus the crust. By the time the closed-circuit lines spread throughout the crust and beyond, they would be quite weak. I'm about to show you further evidence that the magnetic lines exist in actual waves and that they generate the action or mechanism occurring within a magnetic pump, when the magnetic waves compress the attracted atoms within the point of crossover, which is where the pump's mechanism takes place.

What we need to understand most is that all magnetic energy definitely does exist in the form of lines; plus, it exists in a frequency, with the pulse of the Earth as the greatest example. The pulse possesses an actual practical wave form, and there is a pulsing flow of energy radiating along each of the four lines of the diagram.

The magnetic resonant frequencies in examples above and their physical waves resulting in mechanical action are the cause of the hum in any transformer. The hum of a transformer is the result of the physical magnetic waves pulsing out from the windings, thus mechanically vibrating the surrounding materials. Of the two energies within electromagnetic energy, magnetic energy is the component which exists in the form of waves and therefore a pulse within any electrical field, and I'll show you evidence of that in a moment in a magnetic pump.

Solid evidence of the mechanical waves existing in the magnetic component of any transformer can be clearly seen by the fact that a hum only occurs where there is a strong component of magnetic energy. Transformers, especially ballast transformers (used in fluorescent lights), function due to a high degree of magnetic energy. The most obvious example that frequency waves do exist within magnetic energy (greatly due to the magnetic component) is in a one-to-one isolation transformer, which simply separates the supply voltage but at the same time limits the current. Current would be the magnetic component of energy. The one-to-one transformer limits the current but does not alter the voltage by stepping it up or down. It is a current-limiting transformer. This one-to-one current-limiting transformer is best known as a 'safety transformer', which existed in a period of the building industry before earth leakage circuit breakers (ELCBs); these are devices for residual current protection, now nicknamed RCDs (residual current devices).

If we also look at the example of an oscillating transformer, which is used in the circuitry of printed circuit boards, we will notice that current is also the governing mechanism in their functioning. The oscillating magnetic field of the transformer is switching (between a generated field and a collapsing field), thus causing inducement and the movement of electrons. Oscillating transformers are the component which generate the strongest vibration within any electronic components on electronic printed circuit boards. In fact, any electronic component which is a high-current-carrying device is also a component which generates a great deal of vibration, which also occurs in the electrical distribution network.

Out of all the cables used in the electrical distribution network, it's only the cabling which carries the highest levels of power which vibrate. I found this especially true during my forty years in the electrical trade in the most common example of the 'mains cables', which carry high current values whilst feeding high-rise buildings. The high current values of these current-carrying mains cables in high-rise buildings generated nuisance eddy currents. When installing the cabling during construction, if the mains cables were not placed alongside each other in the correct trefoil sequence or phase order, they would suffer the nuisance effect of eddy currents a great deal more. The proper sequence was not required for low-current-carrying cables which fed smaller buildings.

The magnetic pump is the resultant mechanism of the crossover within every sun, and it functions in the following manner. As we now all know, magnetic energy in the form of flux particles flows in the opposite direction to that of electrical current, which is the movement of electrons, generating electricity. Electrical energy flows from the positive polarity of any battery through the device (maybe a light bulb) to the negative polarity of the battery. Remember that the magnetic component will be low during the conductivity of electricity. Magnetic energy flows from the south pole (negative electrical polarity) of a magnet to the north pole (positive electrical polarity). Remember that the electrical component will be low whilst this occurs.

From this, we can clearly understand that there is, in fact, a flow of energy within a magnetic field. We also know that electromagnetic energy does possess or exist in a frequency (according to science), and therefore, so too does the electromagnetic energy when it exists in a high component of the magnetic energy. Magnetic energy very obviously also flows in wave form but in a low frequency (according to my theory).

As the magnetic lines radiate from our sun, there is an outward flow of energy which possesses a small percentage of the electrical component. The magnetic waves continually pulse along the lines in the opposite direction to the electrical energy. The magnetic energy travels towards the centre of the magnetic pump, where the crossover occurs. Where the crossover of these magnetic lines occurs, there is a frequency which possesses magnetic resonance.

The atoms which exist in the identical frequency are gas atoms but primarily hydrogen atoms of gas. When the gas atoms are attracted into the actual point of the crossover where the most reactive part is occurring within that centre point, the magnetic waves are pulsing as they travel down the magnetic lines; they're then compressing the gas atoms (within the crossover) attracted by the magnetic resonance. The pulsing places the gas atoms under a compression wave. The atoms are forced to increase in movement and velocity, thus colliding with greater force (this is what gas atoms are renowned for) and therefore generating greater friction and heat of up to 1,000,000°C within the reactive component of our sun.

Most importantly, remember that the electromagnetic spectrum contains the visible light spectrum in the centre of the electromagnetic line. Another very strong point to make which confirms a frequency within magnetic energy is the fact that science says that every electromagnetic energy on the electromagnetic spectrum is a light frequency, and we all know that every colour of light exists or possesses a frequency. Science says that our sun emits white light, and I am going to tell you that the frequency of the magnetic pump where the crossover occurs exists in a combined frequency which makes up white light. This generated combined frequency consists of every frequency or visible colour and is therefore why the magnetic pump (as our sun) emits the white light.

Because the attracted gas atoms are in resonance with the crossover, this implies that the gas atoms being burnt are also emitting white light. Remember that the visible light spectrum exists in the centre of the electromagnetic spectrum. All electromagnetic energy (which obviously includes the visible light spectrum) travels in photon particles, which are particles of light. The visible light spectrum consists of every visible colour, whilst the remainder of the electromagnetic spectrum consists in every other invisible energy of colour.

The four lines of our sun consists of two lines which have crossed over, and if we were to remove just one of our sun's full lines, we would be looking at one full electromagnetic line, and the visible light spectrum would be positioned in the centre of the full line. Of course, the visible light energies on the EM spectrum are in the centre of these two magnetic lines where the crossover occurs in the reactive centre point, but the crossover of two lines means that there are two visible light spectrums. The gas atoms which are attracted due to resonance exist in that identical frequency and begin gathering in the centre point of the two crossed-over lines. The frequency of our sun has not changed, but the centre area of reactivity (as the magnetic pump which generates white light) has expanded whilst becoming more and more reactive due to the increase of reactive energy.

The combination of every atom (that has been attracted) which are individual molecules of a volatile gas gives off white light as a result of being burnt. I am trying my best to inform you that the mixture of gas being burnt is a combination of frequencies which meld together to make up the white light being emitted by our sun, a little like a number of musical instruments that play a melody. As the reactivity increases within the crossover, so too does the expansion of this area within the crossover point. The more atoms of gas that gather within this expanding area, the greater their level of reactivity, also resulting in increased levels of white light being emitted.

Our sun's magnetic lines had not begun to curve until quite some time after the reactivity had increased. When the lines did eventually curve, they fractured in stages along each line. When the two lines first crossed over, each line was a 'full line' and was unbroken or un-refractured. When our sun, as a magnetic pump, was first generated, it immediately began attracting the resonant gas atoms; and at some stage, both of these reactive frequencies (the frequency of the magnetic pump and the frequency of the attracted atoms) both began to reactively generate white light. The centre point of the crossover possesses the reactive resonance. It also generates the pressure wave which compresses the atoms, combusts them, and emits the white light.

The increasing reactivity also increases the velocity of rotation of the reactive sphere or ball, and the two unbroken lines, which then began functioning as four lines, also began to curve. Whilst the electromagnetic lines were beginning to curve, the reactive centre point where the two lines met—occurring within the crossover of two visible light spectrums—was always expanding as the lines continued spanning out from the centre. It's a little hard for me to explain this mechanism, but in the middle of these two lines where they crossed over, they melded together due to reactivity right in the centre of each line in a very fine point. The centres stayed melded together whilst the energy of each line expanded out. It sort of stretched each line out.

The centre point of the magnetic pump where the reactivity took place expanded over billions of years; thus, the area of the magnetic field and its strength, as the reactivity, also grew. The result was a centre sphere or ball (for want of a better word) that grew in reactivity and mass to what we now have today. The emission of white light occurs due to the increasing reactivity of combusted and accumulated mixture of gas atoms. When the energy of the lines expanded, they also began to curve and refracture, which is where the first eight largest planets were created, but I'll give you more information on this soon. Later, I will be describing how our sun's magnetic lines began to curve and fracture at points along each line in linear stages, which then created each planet.

The following is just a brief point. Each planet first began as merely a 'frequency of light in the shape of a sphere' which existed within the visible light spectrum, and this is why each planet is the colour we see today. The two lines of our sun eventually became segregated in the centre by the reactivity of the ball, eventually becoming four radiating lines, when the increased size of the centre mass occurred.

As stated above, the reactivity grew, attracting a greater influx of atoms. This basic description is what mechanically occurred to the lines and our sun. The details of how the planets were first generated will be described in slow reveals in just a moment, but first, I'd like to expand on my description of eddy currents.

Tornadoes, cyclones, typhoons, and hurricanes are atmospheric vortexes which are generated by eddy currents. Because the open-circuit magnetic lines have the strongest flux density because they do not return to the opposite pole, it is strong evidence that open-circuit lines have very little magnetic flux particle flow, therefore very little electrical energy compared with closed-circuit magnetic energy.

Our Earth has two polarities (north and south), and each of the four lines in the diagram possesses a polarity in a clockwise direction with a north and then a south, a south and then a north. The reason that these lines are so obscure is that two lines crossed over. Each line was a north and south, and this is how our sun and every other sun in our galaxy ended up with a north–south–south–north orientation.

Tornadoes occurring in the northern hemisphere rotate in the opposite direction to a water vortex in the same northern hemisphere. (I'll describe the cause of that in a moment.) Cyclones are atmospheric vortexes occurring in the southern hemisphere. The large vortex as a cyclone (occurring in the southern hemisphere) spins in an opposite direction to a tornado, which is a tight vortex most often occurring in the northern hemisphere. An important fact is that tornadoes, hurricanes, and typhoons occurring in the northern hemisphere spin in an opposite direction to cyclones, which almost commonly occur in the southern hemisphere. The cyclone is a vortex with a diameter which is a great deal larger in circumference than a tornado, which is recorded in the northern hemisphere with a smaller circumference and hurricanes; plus, typhoons also occur in the northern hemisphere with a large circumference.

All the atmospheric vortexes are generated by the magnetic sun lines as nuisance eddy currents when sun lines are projected onto our Earth by either our sun or a neighbouring sun. Remember that a neighbouring sun line which is much larger will point in our direction every twelve months when cyclones occur in the southern hemisphere. A cyclone (of the southern hemisphere) is generated by the large neighbouring sun lines as an atmospheric vortex which is projected onto our Earth every twelve months when cyclones are recorded. A tornado, which has a smaller or tighter vortex, is generated by one of our sun's smaller magnetic lines on king or neap tide days (by the strongest open-circuit lines) but can also be generated by an open-circuit line which exists on either side of the neap tide lines. I am saying that tornadoes occur when king and neap tides are generated, but I'll show you the evidence of this fact soon, which is recorded by the dates and locations of tornadoes. Tornadoes only occur during the day, and this is because the lines of our sun are only pointing on that side of our Earth during the day.

Let me now try my best to explain the technical mechanism of how either of these two vortexes are generated. They're governed by the timed events of our solar system, plus the magnetic polarity of our Earth and the polarity of each magnetic line projected onto our Earth. In general, when two identical magnetic polarities move towards each other, a repelling energy is generated, resulting in a great deal of movement of the magnetic particles. The two identical magnetic polarities in this explanation are **each hemisphere of our Earth** <u>and</u> **the polarity of each line**.

In regard to the north–south and south–north polarity of each line, remember that our sun was originally just two lines (one from a 100 sized sun and one from a 1000 sized sun) which had north and south polarities at each end which eventually became segregated by the reactive sphere or ball. If we look at the lines before they were segregated, we would see a north–south crossover, and this is the way they stayed after segregation. Remember that during power generation, the direction of flow within the copper wire (clockwise or anticlockwise) will depend on which polarity the magnet end moves into the coil of wire. In the coil of wire, the result of movement by magnetic flux particles is generated electricity which can flow in either direction.

When a magnetic line is projected onto our Earth either by a large neighbouring sun or by our sun and reactivity begins, it is the injected magnetic line which will generate the most movement, rather than movement of the solid Earth. This is obviously because our Earth is the most solid, fixed, magnetic object in comparison to the projected line, which simply consists in flexible magnetic energy. Therefore, the flexible incoming line will generate the movement when it spins as an atmospheric vortex.

As mentioned, the large cyclone or a smaller tornado is generated by either a large neighbouring sun line or one of our sun's smaller (king tide or neap tide) lines entering our Earth's atmosphere. The tight circumference of a tornado vortex generally only lasts anywhere from a few minutes to a few hours but never much longer. However, cyclones, hurricanes, and typhoons can last for up to a week, and I'll explain the details of how that momentum takes place in a moment.

Again, when either magnetic line (north or south) enters our Earth's magnetic field in the same polarity, a repelling energy is generated. Because the line is the most flexible, it is the line which will move in a clockwise or anticlockwise direction, depending on polarity, in exactly the same manner or principle as the current moves in the coil of wire when the magnet enters the coil. The polarity of the incoming magnetic line and our Earth will determine which direction the vortex will spin, and the size of the incoming line will also determine the size of the vortex (cyclone, hurricane, typhoon, or tornado). A cyclone or a tornado will never be generated by an incoming line which is in the opposite polarity, and science tells us that the tornado atmospheric vortexes do not occur on the equator, where the neutral energy is positioned. In fact, science says a tornado will not occur within twenty-eight degrees of the equator. Later I'll tell you what occurs when opposite polarities meet on our Earth, generating an attracting energy of an injected line onto our Earth.

As we also now know, a water vortex will spin in the **opposite direction** to an atmospheric vortex in the very same hemisphere. The differing directions of water and air, described above, is because electricity is generated during one phenomenon and magnetic energy being the primary energy in the other phenomenon. Remember that electrical energy and magnetic energy flow in two opposite directions. Spinning eddy currents within magnetic energy occur during atmospheric vortexes, and water vortexes occur under the mechanism of eddy currents within electrical energy. Electricity is the movement of the negative electrons, but also remember that it is still electromagnetic energy, which means it will possess both magnetic and electrical energies. This is a little ambiguous but is the best I can do to explain the differing directions of rotation between water and atmospheric vortexes.

I also briefly mentioned the timed events. A cyclone will only occur due to our Earth being on a tilting axis, causing the area of our Earth to be facing that incoming neighbouring line. The energy of a cyclone is generated at night but continues to spin during the day due to momentum. A tornado will only occur during the day when our Earth is facing the magnetic lines of our sun. Tornadoes only occur for a relatively short period in comparison to cyclones, and I'll show you more evidence about these two different-sized atmospheric vortexes soon.

How Our Planets Were Formed

The following are basic numerical facts of how all our planets were formed after being generated on the lines. I'm about to describe how each planet was first formed, starting with a recap.

One metallic planet was formed on each of the four lines, and therefore, there are four metallic planets. One giant gas planet was formed on each of the four lines, and therefore, there are four giant gas planets. There was only one early dwarf planet (the largest today), formed on each of the four lines; and therefore, there are four early and largest dwarf planets. So four lines and three different planets were formed on each line of the diagram. The other dwarf planets were also formed on the lines, but I will tell you about those later.

The sun lines were in pairs by size due to the lines having been formed by just two neighbouring lines, which came from suns 100 and 1,000 times bigger. After the reactive centre segregated the lines, they were then in pairs by size, and the planets were also formed as planet pairs, also by size. The manner in which each planet was actually generated also caused the planets to appear in a line of colours of a partial visible spectrum.

The generation of the planets and my theory which describes the colour of each planet (existing in a visible spectrum) are a little technical for now, so I will reveal both of those theories after we get to know these magnetic lines a little better in a few more chapters. I'll show you some reveals now, backed by a few facts of science, which will help support my theories and will also help us understand this slightly technical chapter. The following is a description of the planets in a general and numerical manner.

Each planet grew within the magnetic framework of its paternal magnetic line, and later, I will describe the magnetic lines in detail, which will give you a greater insight into this magnetic framework and how it was able to hold each planet within the framework as they rotated around our sun. **Each planet was growing on each paternal line, and the following is what occurred next.** The following will help you see other similar facts, which I refer to as 'further mechanics of the four lines'.

Before we look at how each planet began as merely a frequency of light, I need to take you through some more steps in stages of information and facts that will help you see how it all occurred, beginning with the consideration of a bamboo cane as it whips through the air. This will actually help us clearly see why metallic planets are the first in that chart of ascending planets.

Curving caused each line to fracture at linear points along each line but in stages. Remember that we're looking at it mechanically for now, with a deeper technical description later about the actual generation of the fractures as 'planet points' and 'light frequencies'. I use the term 'generation' because that is actually what magnetically occurred to the magnetic energy of each line.

When striking a cane through the air, the cane would begin to curve first at the hand, and the curve travelled out along the length; so too did each magnetic solar system line of the diagram. The curve of each magnetic line travelled out in linear form from where each line was affixed (at the reactive centre mass); there was a first point of fracture which was at a 'ratio distance' away from the centre mass. This first fracture point occurred at a distance relative to each line's length and flux density (strength), which will eventually have a mathematical formula. *(I'd like to identify this formula now as, 'the magnetic ratio, but I haven't gotten far enough into my investigation yet to tell you what the exact formula is.)*

As rotation velocity increased, so too did the force against the front face of each line until the first fracture occurred to each of the four lines. This first fracture (in calculation of the magnetic ratio) is where each metallic planet point was first generated. As the reactivity increased, so too did the rotation velocity of our sun, and so too did the force against the front face, thus further curving along each of the magnetic lines until the next stage (calculated by the magnetic ratio) **until the next planet fracture point occurred, which was a giant gas planet fracture point** on each of the four lines. As the linear curving continued (caused by increasing velocity), so too did the fracturing in the magnetic ratio; and eventually, curving occurred all the way to the end of each magnetic line, where each dwarf planet fracture point was eventually generated and grew.

The Mechanics of the Two Original Magnetic Lines

I can tell you that when our sun was formed by sun lines 100 and 1,000 times bigger than ours, our sun then existed with two different-sized lines, radiating from the reactive centre mass. As rotation velocity increased, the largest line of our sun (formed by the 1,000 line) began curving first primarily due to the length but also due to its larger circumference, thus the increase in its resistive force. Each 'pair size of the lines' was positioned on opposite sides of each other when we visualise our sun's centre mass, with four lines radiating out.

In the planet chart, there are undeniably two large metallic planets and two smaller metallic planets. They were formed by two different-sized sun lines, but today the sun lines are almost the same size, which is reflected by the size of the pairs of king tides where they do not have a great deal of difference.

I can also tell you that because the larger line (which was eventually divided by the centre mass and now functions as two lines) began curving first and this line created the first two fracture points which generated our Earth and Venus as the larger metallic planets. These first two fractures could be calculated in distance from the centre mass by using the equation from the magnetic ratio, where these first two larger metallic planets were generated. Which one of the larger metallic planets (either Venus or Earth) were **the first to be generated** has not been possible for me to theorise (just yet). Both of the largest metallic planet fractures (on the largest line) couldn't have possibly have occurred simultaneously, but there may actually be a very slim chance that they did (more work for another day).

Very importantly, each first planet point occurred on each line in calculated stages at the magnetic ratio out from the centre mass according to their mechanical structure, and I refer to the distance that this occurred (in lay terms) as 'equal distancing'. By using the magnetic ratio (when eventually discovered), we could also determine where the next two planet fractures would have appeared. Each fracture point continued along each line in the magnetic ratio distance to where each planet point was generated and then formed, which was also at equal distances. The equal distances (between planet fractures or planet points) may one day all be possible to calculate by using the magnetic ratio.

These two facts of 'lines at differing lengths' and 'equal distancing' by the magnetic ratio begin at a distance out from the centre mass and are the cause which gave each of the first metallic planets their positions when and where each metallic fracture point occurred. The equal distances (calculated by the magnetic ratio) also apply directly to rainbow colours, which exist in equal distances. The distances between colours within rainbows is also the magnetic ratio in action.

I must explain that, in the early years when the sun lines began to fracture and produce planet points, the lines were in two different definite sizes. There would have been two large lines that fractured first due to an increased resistive force on the front of each line. In the early years, the size of the lines would have been very prominent, and the smaller stockier line would have fractured second.

<u>The reactive spherical centre mass is the source of the driving force generating our sun's rotation.</u> Later, I'll describe the details of how I discovered the rotation frequency of our entire solar system, which is every four years. In fact, it helps to somewhat visualise our sun and each of the four lines as (more or less) an electric motor with four long flaps hanging off it.

Under similar laws, the revolutions per minute is the rotation velocity of an electric motor, which is governed by the number of poles of the motor. This electrical law about motors regarding the 'number of poles' is also a direct relationship determining the rotation velocity of our solar system, and later, I'll show you where the law of poles applies to the calculation of our sun's period of rotation. Our sun has four magnetic poles, and the energy of these poles radiates out as each of the four lines.

You also know that they were generated when the crossover of the two original lines occurred, and the segregation and division of the original two lines by the centre spherical mass also occurred. The original two lines now function as four lines which were always magnetically just two lines but segregated by the centre mass and now actually function as four separate lines when they generate each of the four king tides after inducing magnetic energy into our moon's normal gravitational energy. Our moon's normal energy is gravitational and could not possibly be magnetic because magnetic energy is much stronger than gravitational energy. I am saying that the magnetic energy of each king tide line generates a king tide and so too for each of the neap tides. This is how it is possible that the four lines can generate a greater tide than the normal gravitational energy, remembering that normal gravitational energy is weaker than magnetic energy.

Those two original lines are ever present today and are still also magnetically functioning as 'pairs of lines', seen by the evidence in the generation of the king tide pair sizes each time they induce the moon's normal gravitational energy with magnetic energy. The king pairs occur twice during each orbit as the four strongest king tides but, as stated, in pairs by size. Further prima facie mechanical evidence (of two different-sized original lines) can also be seen by the existence of the planet pair sizes, and that evidence will be clearly supported later in the chapter titled '**The Frequency Generation of Each Planet**'.

There are also two small and two larger giant gas planets. I have identified the two line sizes in the simplest of terms as the 'two 100 times bigger lines' and the 'two 1,000 times bigger lines'. The shortest of the two original lines (created by the 100 line) was eventually divided by the reactive centre spherical mass after it began to curve as two separate magnetically functioning lines, followed by fracturing (at the magnetic ratio planet points), which were generated as the first steps, at equal distances from the centre spherical mass as the second pair or smaller pair of metallic fractures. And so after the larger line fractured, the smaller and shorter line's first fractures also occurred (generating the two smaller metallic planets) at the magnetic ratio and 'equal distances' from the centre spherical mass but at closer distances to the centre mass than the two larger metallic fractures due to the length of the lines and the law of the magnetic ratio distance (https://goo.gl/images/46VoDG).

What I am trying my best to say is that the larger line fractured first but the larger fractures occurred at a longer distance out from the spherical mass than the two smaller lines. This was because the longer line had fractures which were further out from the centre mass or sphere. This meant that the first two metallic planets which are larger were formed at a longer distance away from the spherical mass than the two metallic planets on the shorter line. (Remember that the fractures occur at distances which one day will be calculated by the magnetic ratio distance.)

The larger line's first fractures occurred at distances further out from the centre spherical mass than the shorter line. From those two simple mechanical descriptions (larger line metallic fractures and shorter line metallic fractures), we can clearly see why the four metallic planets now exist today in two different-sized planets that I refer to as 'planet pairs by size'. When we look at the ascending line chart, we will notice that Mercury (small metallic) is closest to the reactive centre mass; next is Venus (large metallic) and then Earth (large) and Mars (small), but Venus and Earth are the largest, with the two smaller planets in positions before and after them.

The largest two metallic planets, which are Venus and Earth, are side by side and in the centre positions of metallic planets. *Theoretically, the pair of smaller metallic planets should be first in the ascending planet line and should not be divided by the two larger metallic planets on the ascending planet line chart.* The positions of the metallic pair sizes today should somewhat resemble the positions of each giant gas pair, which are side by side.

The largest planet, which is Jupiter, is first in the line of giant gas planets, followed by Saturn. This means that the large pair of giant gas planets are first in the ascending line, followed by the smaller pair of giant gas planets. *The largest giant gas pair should theoretically be last in the ascending line according to my theory of the planet positions, and the smaller pair of planets should be first, but they're not.*

According to my theory, the smaller giant gas pair fractures occurred at a distance closer to the centre spherical mass. The smaller giant gas pair should therefore be in closer positions to the centre spherical mass before the larger giant gas pair, but they're not. *The reason for this is a bit of a mystery for me so far, and I can only suggest that the larger giant gas planets may contain a denser material and therefore be drawn closer to our sun due to the sun's gravity. It would make more sense for this to be the fact, but due to the two larger giant gas being closer to the four metallic planets, how this actually occurred is an even greater mystery for me.*

This fact (of where all metallic and giant gas planet positions exist on an ascending line chart) shows a gravitational force by our sun as the source involving the final ascending planet positions. I honestly think today that every planet's orbit period is due to the frequency of the closed-circuit lines which radiate out from our sun, but it's another week's work to describe it in full.

All metallic planet positions may have been caused by the frequency of the closer metallic planets (as they exist today in an ascending line) having resulted probably in an accumulation of a greater metallic density of the attracted atoms (then molecules) which finally formed as an actual closer metallic planet. *I hope I described that properly. I'd hate to have to edit that again. (My dyslexia can be painful).*

Later in the chapter on the frequency generation of all planets, I will describe how each planet began as merely a frequency of light which emitted a colour that exists in each planet today caused by its individual light frequency. Remember that every electromagnetic energy on the electromagnetic spectrum is a frequency of light and exists in a colour, especially the colours on the visible light spectrum.

After each line began to curve, it generated metallic fractures first, thus metallic frequencies, and the curving force continued the length of each line. And each line (just like the cane) continued to curve but in stages to where each giant gas planet point then occurred at specific positions along each line due to the magnetic ratio distance. Each line continued increasing in rotation velocity, along with the consequential linear curving, all the way to the end of each line, where the four (earliest) dwarf planet fractures were generated also under the law of magnetic ratio distance. There were now three planet fracture points along each line in equal distances according to the MRD.

Each planet fracture began under much the same mechanism when the two suns' lines generated each magnetic pump, but our sun possessed only one point of magnetic resonance. Whereas each planet point fracture (also possessing magnetic resonance) had two points that were generated as the field of magnetic resonance. Unlike the frequency of our sun which attracted gas atoms, the frequency of each planetary light point (as the beginning of each planet) possessed their own magnetic resonance as planet frequencies which possessed an extremely low micro-magnetic amount of energy which grew to eventually be able to attract particles that were in resonance with the materials that make up each planet today.

As a recap, each line began curving at the closest point to the driving centre spherical mass; and under the law of magnetic ratio distance, fractures occurred, thus generating the four metallic planet fracture points. Then as curving continued, the next four planet points appeared under the magnetic ratio distance (MRD), where each giant gas point was then generated, followed by the four early but largest dwarf planet points.

Science today actually records a total of eleven dwarf planets. There are the four early but largest dwarfs, plus seven other dwarfs that I refer to here as the seven micro-dwarfs. A total of eleven dwarfs are recorded today by science, and I theorise that they were originally composed of two groups of eight.

The first group of eight dwarfs consisted of the four early or largest dwarfs, plus four other dwarfs which were from the remaining seven micro-dwarfs. The next group of eight dwarfs composed of three from the remaining seven micro-dwarfs, plus five more micro-dwarfs which are presumed missing and orbiting in another solar system today. Science today describes some dwarfs as micro-sized dwarf planets which orbit in and out of neighbouring systems and are not permanent in our solar system.

Remember that every sun exists in the identical frequency; therefore, every other sun's solar system frequency will be a familiar energy to all dwarf planets which may be entering into neighbouring systems. This theory of familiar frequencies will be logical, making it possible for micro-dwarf planets to exit our solar system and re-enter into neighbouring systems, which is just another cause of resonant frequency energies. This is a point of interest that I will logically prove to you beyond any doubt soon. So the other five micro-dwarfs which are missing make up a total of sixteen dwarfs, with eleven accountable and the remaining having left our solar system and presumably now orbiting in neighbouring solar systems.

Further Mechanical and Physical Properties of Our Sun's Lines

The largest planet pairs from each group (two large metallic and two large giant gas) were generated on the largest line, which was also obviously larger in circumference. There was also another magnetic line, which was a hairline that projected out from the end of each of the four lines, where the first group of eight dwarfs were generated after that line also eventually curved.

As stated, those first eight dwarf planets included the four early or largest dwarf planets, plus four of the seven micro-dwarfs. If we look again at the ascending line chart, at the recorded size of the first four early dwarfs, and at the recorded size of the seven micro-dwarfs, we will notice a substantial difference in the overall size of all dwarf planets, especially the largeness in size of the four early dwarfs compared with the other seven micro-dwarfs. This difference was the same for the first eight original planets, metallic planets compared with giant gas planets. Quite simply (just like the original eight planets), all dwarfs exist in two different sizes, which meant that the hairline also created two different sizes, just as the line that the first eight planets were generated on.

The first four early or largest dwarfs are a great deal larger than the remaining seven micro-dwarfs. This same comparison can be made about the eight largest planets (i.e. the metallic planets are much smaller than the giant gas planets). Jupiter the giant gas planet is around 395 times bigger than our Earth which is a metallic.

To properly understand why the two different-sized dwarfs exist, we need only to realise that the first largest original eight planets were also in two different sizes (i.e. metallic planets were smaller than giant gas planets). So there are the first eight dwarfs (which includes the four largest dwarfs), plus three other micro-dwarfs and a total of eleven today. There were also five other missing micro-dwarfs, making the actual total of sixteen dwarfs. This final group of eight micro-dwarfs (including the five that have exited and are now missing) was generated from fractures along another 'finer hairline', and as stated, five of those last group of eight dwarfs have exited into another neighbouring solar system or systems.

Further Mechanics of the Lines

The eight largest original planets were the first group of eight planets and were generated out of the very first eight largest fracture points—Mercury, Venus, Earth, Mars, Jupiter, Saturn, Uranus, and Neptune. The first four metallic planets were the smallest of the eight largest planets listed.

I'll show you the evidence in a moment that the solar system lines which created the metallic planets and then the giant gas actually existed in a shape which was much like a witch's hat but with the smallest end of the hat affixed to the centre sphere or ball mass. This reveal is a slightly confusing theory in *The Holy Grail of Science* but will become very clear soon in the chapter on the frequency generation of all the planets. The description about the shape of a witch's hat is about the hardest to understand out of all reveals here, but it will also be described in detail later, where it will all make greater logical sense. *So if you need to go back and read the chapter again, I advise a little break first.*

The next group of eight planets consisted of the first eight dwarfs and were created from the hairlines, which must also have logically exited out of the largest lines which created the eight largest planets. The next finer hairlines (also in the shape of a witch's hat) must have also had its smaller end attached to the end of the previous largest line. This theory of cone-shaped magnetic lines also applied to the finer hairlines which generated the last group of other eight micro-dwarfs, and I'm trying to say that four fine hairlines were also theoretically exiting out of the hairlines in cone shapes, with the smaller end of the cone attached to the largest end of the previous hairlines.

One last fact for planets that I need to reveal is that, recently, I was told that there were once about thirty to forty micro-dwarf planets that orbited our solar system (recorded by modern science); and at some stage, they have left, presumably now orbiting in other solar systems. I'm about to show you how that was possible in a very basic but simple description.

As I said, all the other missing planets have also existed into neighbouring solar systems. The following is strong evidence that every planet was formed on a line which I now refer to as magnetic frames, followed by a clearer description of the actual structure of these lines. The strongest evidence I can give is that each planet grew within a magnetic frame and was (more or less) held in place by the framework of magnetic energy whilst rotating around our sun. Each planet grew on its paternal line to a particular size or mass and then left each line behind. Each planet grew within each frame, with the greatest evidence being given by supporting scientific facts about our Earth throughout its history.

If each metallic planet was growing on a line at equal ratios from the centre mass, then each metallic planet would have been on a collision course with the next planet from it. Have you got your hat again?

I was working on the theory that each planet grew on a paternal line and eventually left each line, headed for another planet, when I stumbled on an actual scientific fact that says Mars is known to be in a D shape, and our moon is said to have once been a chunk of Mars that broke off. This meant that, in accordance with my theory, Mars collided with our Earth, and a chunk of Mars broke off which became our moon. I'll tell you more about that theory after we look at other facts of greater interest.

Our Earth was once a one land mass continent that science identified as Gondwana. I have described the growth of our Earth as a planet that was generated from a fracture and a globe which then formed within a frame of magnetic energy on just one line. I can tell you that our Earth was forming and growing in mass within the frame of its paternal line and also that, during this period of growth, it did not spin every twenty-four hours. *(I use the term 'spin' because the term 'rotate' is used later in reference to another mechanism.)* In fact, during this period of growth (within the energy of its line), our Earth did not spin whatsoever. I refer to this as our 'stagnant Earth' period. The fact that our Earth did not spin is the reason Gondwana was able to grow as a one land mass.

The best way to show our Earth was not spinning and becoming (more or less) stagnant is to describe it as a sphere which was (more or less) facing into the wind and accumulating particles on the front face of the globe or the front face of the stagnant Earth. All this dust and larger particles eventually grew into Gondwana, the one land mass continent. *(I'll show you the supporting evidence by other facts of science in a moment.)*

Whilst our Earth was growing within the magnetic frame, it was therefore rotating around our sun (within the frame of the line) at the same rotating velocity as our entire solar system's own rotation period (every four years). Our Earth's rotation velocity around our sun or frequency was therefore identical to our solar system and the rotation of our sun's centre spherical mass, which is the source of the driving force.

I can tell you that one rotation of our sun and its four lines took (and still takes) four years, which is, of course, a quarter of our Earth's orbiting frequency today. It may help to visualise our sun as (more or less) an electric motor with four arms affixed to it and our Earth attached to one of these arms, rotating along with the frequency of the motor (for want of a better description). I use the term 'rotation frequency' in regard to our sun and solar system because 'orbiting' is a mechanism of each planet today. Every planet during this time was affixed to the arms or lines and did not begin orbiting until they had left the magnetic lines.

Whilst our Earth grew within the magnetic framework, I refer to our Earth's frequency as its 'rotation period' (around our sun) and not its orbit period because 'orbiting' is the later mechanism of our Earth after it left the paternal line, thus beginning its own orbit pattern. I refer to some planets eventually leaving their paternal line as being **ejected off** their line, but ejection did not occur to every planet, and that fact has very strong supporting evidence, which

I'll show you very soon. *I identify the ejection of planets as* **magnetic planetary ejection**.

Again, the rotation period and said velocity of stagnant Earth as Gondwana was every four years. In fact, as previously mentioned, four years is the rotation period of absolutely every one of the 400 billion suns throughout our galaxy. Remember how extremely synchronised our 500-year event is, plus every other synchronised event, and also that 4-year rotations evenly equate to 500 years. This rotation period and velocity of the centre mass of four years was a quarter of what now occurs to our Earth. It was therefore very slow in comparison to the period after our Earth left its paternal line and began its own orbital pattern.

Every planet was eventually destined to grow to such a size and mass that each planet's line was no longer able to support it. Not every planet was ejected. Those that were ejected had amassed quicker, or their attracted compounds, which each planet consists of, weighed more (due to the molecular structure of attracted materials), and these heavier planets were the first to be ejected. In fact, Mars was one of the first metallic planets to be ejected off its paternal line, and it then began to orbit in its own pattern, at a rate which was much higher in speed than when it was on the line.

Mars was ejected off its paternal line and was no longer rotating around our sun at the rate of our sun and its solar system as it had done whilst growing on its line. When Mars was ejected and began its own new orbit pattern, our Earth was then in its oncoming path and had not yet left its line, rotating around our sun at a very slow rate. Remember that Mars grew on a smaller line, and our Earth grew on a large line, and the rotation order of the lines begins with a small line, a large line, a small line, and then large line.

Our Earth was the next planet around on the oncoming path of the new Mars orbit. When Mars began its own orbital pattern, it struck our Earth from behind in one massive ocean that covered the majority of the planet and then punched our Earth off its paternal line. Science recently reported that the same materials exist in Mars as in our moon and also that Mars is in a D shape. I say that, when Mars punched our Earth off its paternal line, a large chunk of Mars shattered off that somehow became our moon.

I was working on this theory of Mars having punched our Earth off the line for quite a long time before science announced the chunk of Mars as our moon, but it certainly helped me confirm a few things in a logical manner. I can prove to you that I was working on this theory about Mars and our Earth well before science announced their discovery. Every time I have worked on my book over the past ten years, I have saved each copy with a timestamp, which will prove that I theorised this first before science announced it.

As our Earth rotated around our sun whilst being fixed to the magnetic line (held in by the magnetic frame), it steadily grew, but it did not rotate but steadily accumulated particles on its front face, identified as the one land mass of Gondwana. Our Earth as Gondwana had one massive ocean, which was identified as the Panthalassa; and when Mars punched our Earth from behind, it hit Earth in the middle of that ocean because the one land mass was at the front of the globe, gathering particles, and the ocean was behind.

Other evidence of this is seen today in reports by science of a split in the one soft land mass which occurred like a cracked egg to our Earth between the continents, causing a massive outpouring of lava. When our Earth was struck from behind, it was incredibly soft in comparison to today. Mars struck our Earth and pushed all the lava to the front of the globe and out of the softer crust, just as science describes it. Mars is recorded as having water on it today, along with our moon. Most likely, the water came from the ocean when Mars punched our Earth.

As mentioned, not every planet was ejected off its paternal line, and there is another planet that science discovered recently that has a hexagon-shaped moon orbiting around it. That hexagon moon was a chunk that shattered off a planet with a hexagon molecular structure, and I'll name that planet later and describe how that structure was generated in some more reveals.

After our Earth was punched off its paternal line, I refer to the result as Earth having then been pool-balled through space. When our Earth was pool-balled through space, it left its warm atmosphere behind and entered into super-sub-zero temperatures, which today are estimated (for the area around our Earth) at -300°C. Remember that -300°C is heated by our sun. I can tell you that life on Gondwana began supposedly from just four species, and I'll tell you later where they came from as well.

The pool-balling into super-sub-zero temperatures caused the first ice age when that stagnant Earth was snap frozen along with nearly every dinosaur on it. During the first ice age, science says that some of the dinosaurs on Gondwana (stagnant Earth) survived. I can't remember all the facts, but the dinosaurs that did survive were very obviously in the outer regions of that enormous ocean or in caves when our Earth was struck by Mars. The pool-balling occurred whilst those dinosaurs were 'out of the wind' (so to speak) whilst our Earth underwent this pool-balling through super-sub-zero temperatures.

After our Earth was punched off the paternal line, it eventually slowed down and began its own orbital pattern of 365 days, thus its new orbiting frequency from that point on. Our new orbiting Earth also began to spin every 24 hours, and both the new 365-day orbiting frequency and the spin every 24 hours gave rise to the next piece of supporting evidence.

The new orbit and rotation was first occurring in an elliptical pattern due to our Earth being lopsided, beginning to spin out of balance, with Gondwana sitting on one side of the globe as a mass, generating this out of balance. Remember that it is movement within a magnetic field which generates electricity and shifts the electromagnetic energy on those scales. As a stagnant Earth (growing on the line), it did not really move very fast through space and did not spin every twenty-four hours. So this stagnation did not generate a great deal of the other electrical component on those electromagnetic scales.

To generate the component of electrical energy and then shift the scales from the primary energy of magnetic, we simply move any object within a magnetic field. Most importantly, electricity is generated when this movement causes cutting of flux lines. I'll describe the finer details and show evidence later that proves beyond any doubt that evolution is directly dependent on the amount of the electrical component (tipping of the scales) of energy.

Whilst our Earth was stagnant, it was fixed on its paternal line and rotating along with the magnetic field of our sun which was projected out into our solar system. It moved along with the magnetic field whilst it rotated around the sun, fixed to the paternal line, which meant that it was not cutting flux. When our Earth was pool-balled by Mars and it left the paternal line, it also began orbiting throughout the strong magnetic field which exists within our solar system and the many magnetic lines which our sun projects across our solar system. The four species of dinosaurs which roamed Gondwana were almost wiped out by the pool-balling and were recorded as having very low intelligence compared with the species that followed.

The new orbit pattern (with an increased frequency spinning every twenty-four hours in an elliptical pattern) and the fact that this new orbit was now within a stronger magnetic field (throughout our entire solar system) constituted a greater percentage of the electrical component. With a great deal of the electrical component generated (especially during the stage of pool-balling), the description above helps us understand why I identify Gondwana as a stagnant Earth because it was exactly that in regard to its extremely low component of electrical energy. The evidence of a greater component of electrical energy caused by orbiting through a strong magnetic energy field occurred during the period that I identify as our 'early Earth', which was after the pool-balling.

After our Earth was pool-balled through space (when struck by Mars), it then began to orbit in a new 365-day cycle; our Earth also began to rotate every 24 hours but in an elliptical pattern caused by the lopsided Gondwana. This, in turn, caused a gyro effect which broke up Gondwana into the first continents. It is only partial evidence (I will give you more later), but I can tell you that the sudden appearance of the first species of spider is recorded by science to occur around this period when the increase in the electrical component occurred.

The first species of spiders suddenly appeared during this time when Gondwana separated into many smaller continents. The evolution of the spider is described as an extremely sudden evolvement from a six-legged Scorpion like creature into the very first eight-legged skinny-bodied spider. Science records the sudden evolution of spiders to occur during the period just before our Earth broke up into the many new continents, when as I described our Earth was beginning a new 365-day orbit and 24-hour elliptical rotation after being pool-balled, consequentially increasing movement within the NEW magnetic field.

Our stagnant Earth as Gondwana was very, very soft in its cores and composition, especially in its outer core, due to having only just been created and being (more or less) like a newborn baby. This is why I initially refer to it as the paternal line within which our Earth first grew as Gondwana or stagnant Earth. (My apologies again for such lay terms.) It was the new orbit frequency but mainly the new spinning in an elliptical pattern every twenty-four hours which generated a gyroscopic force that caused Gondwana to break up into the first continents, almost as they are today. (After Gondwana broke up, there was actually 'more carving up' carried out to the shape of those continents, which I will describe in later chapters.) So that you may feel the strong effect of a gyroscopic force and discern how powerful the gyro force was for our stagnant Earth placed on Gondwana (as an orbiting, out-of-balance planet), you only need to spin a pushbike wheel while holding it with both hands and then roll the wheel around whilst holding the axel in your hands.

Let me also mention just briefly here the recorded sudden evolution of another recently discovered creature in 2010 on our Earth, the cane toad here in Australia. It is now documented that the cane toad has made the very long migration into the far north of Australia (away from Queensland, where it was introduced) into Kakadu National Park. The toads that are now entering into Kakadu and are recorded with longer legs obviously grew through a sudden increase in intelligence and sudden evolution. The most important fact about the toads is that it occurred within a very short period of about fifty years at a great rate of speed when the planet alignment occurred, when there was an increase in the strength of the magnetic field.

I must now add that, during the pool-balling period of our early Earth, it would have also been moving very fast through space until it slowed down, and maybe this higher speed added to the generation of the electrical component which, in turn, caused the sudden evolution of the first spider species. Also, pool-balling and spinning 24/7 meant that our stagnant Earth with one land mass was out of balance and would have (more or less) been spinning in an elliptical pattern which (at the time) would have also increased Earth's movement within the magnetic field. Remember the facts which directly affect the level of electricity generated, beginning with movement within a magnetic field or cutting lines of flux. I will also show you further evidence later of the strength of the magnetic field during this period within our solar system. In short, the magnetic field was very intense.

The early Earth's stagnant rotation was a rotation frequency equal to our sun's rotation period, every four years, at a quarter of the velocity today, which theoretically also meant that the four species would have been very unevolved due to a low component of electrical energy. I can tell you irrefutably that the sudden increase in the electrical component was the direct cause of the Scorpion like creature very quickly evolving into the very first skinny-bodied spider. Again, science recorded every Gondwana species to have low intelligence. The next species (which were also in much greater numbers) were orbiting at the new higher frequency and spinning 24/7 during the period of orbit through a stronger magnetic field, thus increased component of electricity, and this was also the cause of a higher intelligence in the next varied number of dinosaur species.

When our Earth slowed down from having been pool-balled, the great number of dinosaurs which were killed during the pool-balling (when they were snap frozen) would have eventually thawed out and began rotting into bacteria, but where did the bacteria come from? Recently, science discovered lots of bacteria on the windows of a space shuttle after its return from space. Our Earth was struck by Mars, which was also laden with bacteria that obviously came from space. Science now says that space is obviously inhabited by large sums of bacteria.

When our Earth was travelling through space during the pool-balling, the atmosphere of our Earth was removed, and it was then unprotected whilst being subjected to the atmosphere of space, which is full of bacteria. You may have seen in social media a number of peculiar species that have recently been found in our oceans. I believe this is due to the electrolytes in our oceans which help the bacteria evolve and come to our Earth inside meteorites. I'll later describe evolution with greater detail under the chapter on evolution of the species.

Identical Speeds whilst under a Vacuum

Previously, I mentioned that light and electricity were announced as travelling at identical speeds whilst under a vacuum. The atmosphere where light is being transmitted in fibre optics is, of course, Earth's atmosphere. A fibre optics cable cannot hold a vacuum, and they discovered that light transmission is only roughly about 100 times faster than electrical transmission, which is surprisingly low.

Light is said to be made up of photon particles. In lay terms, science says that atoms are particles, and so too is anything up to about the size of a breadcrumb, including dust and anything in between. For my own perspective on light particles, I visualise photons as (more or less) specks of dust, and this magnified visualisation helps me realise the amazing fact that light has a physical surface area or mass and will suffer resistive values due to its particles having that surface area. The resistance against the front face of a photon causes it to slow in velocity in Earth's atmosphere.

Light travels faster when under a vacuum because a vacuum has less resistive values. A vacuum is the condition of the atmosphere in space, and we'll get back to that in a moment. The fact that light travels in photons and that their mass decreases the actual velocity of light shows us that light undoubtedly has properties which possess surface area and mass.

I'm sure that you will agree that magnetic energy has far greater values of mass than light energy; therefore, magnetic lines will suffer a greater value of a resistive force. Science says that every electromagnetic energy (including magnetic energy) travels by way of photon particles at the speed of light when in space. The two different conditions compared above, in regard to the velocity at which light will travel on our Earth and in space, help us clearly see that everything will suffer a resistive force to varying degrees in both atmospheres. Especially and most importantly, each of the four lines will suffer a resistive force against the front face as they rotate.

For the next step, we need to look at the following. Science says that light and electricity are irrefutably directly related because they travel at identical velocities when under a vacuum, and there is a reason why they travel at identical velocities whilst under a vacuum. The fact that every frequency of electromagnetic energy travels in varying waves of light is the foundation of how each planet first began from merely a light frequency which grew ever so slowly after billions of years. Each planet began as merely a 'frequency of light' which also possessed an extremely minimal level of magnetic resonance that grew in its level of energy over millions or billions of years.

I'll get back to the finer technical facts of how each planet first began later, but for now, let's just look at what happened to each planet after they had already been generated (within the magnetic frame of each line) and eventually formed as a mature mass before we look at the most technical reveals.

The following description is the actual electromagnetic structure of all magnetic lines. Just for now, I'm still describing the basic mechanical structure of magnetic lines as '**the mechanics of the two original magnetic lines**'.

Again, **every energy** that exists on the EM spectrum is a frequency of light, and that includes the visible light spectrum of seven variable frequencies which is the middle of the EMS. Let me just show you again the basics of how those thirty to forty planets came to be. The first largest eight planets were generated from the frequencies of light which are in the middle of the EM spectrum. Those are the visible light frequencies. After many years of reactivity by the magnetic pump, the lines began to curve, fracture, and then expand out.

ALL the other planets (thirty or more) were generated from all the other frequencies outside the visible light spectrum. Those frequencies are radio waves through to gamma waves, including infrared and ultraviolet waves. The range of the frequencies outside the visible light frequencies also exists in a number of bands which also generated a number of planet point fractures after each of those bands/frequencies also fractured.

As a good example, I'll describe just the range of bands that exist within radio wave frequencies. Radio wave frequencies actually exist in nine different bands of radio waves, and each band generated a planet fracture point. See https://www.livescience.com/50399-radio-waves.html. Please scroll down to see the list of nine radio bands. The nine bands of radio waves are just one example, and if the link disappears, just look up the nine bands of radio waves.

The two frequencies immediately outside the visible light spectrum are the two invisible frequencies which are infrared and ultraviolet. These first two invisible frequencies are normally referred to by science as the common invisible frequencies, and they also exist in different bands of invisible frequencies of light. There are three separate bands of infrared light (https://en.wikipedia.org/wiki.Far infrared). There are also three separate bandwidths within ultraviolet light; please scroll down to just below the third paragraph (https://www.livescience.com/5032-what-is-ultraviolet-light.html).

Allow me to show you where invisible frequencies apply in our solar system. Recently, I read that science, some time ago, sent a signal out to our moon, and it bounced back within 2.7 seconds. They sent another signal out into a region of space where the signal should not bounce back, but the signal did return after forty seconds, and now they are baffled about the structure that it bounced back off and returned from.

Science also describes a black hole with a solid centre spherical mass as if it were a planet. I can tell you that black holes (in lay terms) are the invisible planets which were generated and then formed from the band of invisible frequencies of light. Some of the waves of invisible frequencies are quite large, but I've not yet theorised the size of all invisible planets as black holes.

I also just can't quite yet figure out why the dwarf planets were created from all the other bands of invisible frequencies—gamma waves, X-ray waves, microwaves, and radio waves—yet the dwarf planets are quite visible. (It may, of course, have to do with the resonant particles/materials which were attracted to form each dwarf.) The above theory describes invisible frequencies which are each broken up into a number of different bands and therefore also created a number of dwarf planets. (Those theories are somewhat ambiguous so far, that is, there must be more than forty planets.)

Again, the gigantic planet recently announced in September 2018 at the outskirts of our solar system does clearly belong to that neighbouring solar system. It could be a habitable planet up to 1,000 times bigger than our Earth. That planet is also described as having a moon that science refers to as a planet revolving around it. All moons are recorded as planets, but it isn't a term often used in reference to a moon.

Every magnetic line exists in the identical combination of light frequencies. When broken or refractured along the full length whilst curving, the line WILL generate all the frequencies within it as points of planet frequencies. I'll show you now my theory of what occurs to the lines whilst refracturing. **This explanation begins with the description of the actual electromagnetic structure of all magnetic lines.** The following is simple to follow and actually is quite significant. It is, however, just the basics and will be followed with a finer description later.

All the evidence in the description of planets began to show me that somehow every planet was formed on each line, but at the time, I just couldn't figure out exactly how. The steps to creating a planet were discovered when I began asking myself, 'How did a magnetic line create a spherical-shaped planet?' With the long list of fours, especially that the planets existed in groups of fours, I began to contemplate how it was possible that each planet was formed on each magnetic line in the shape of a sphere.

I kept looking at four lines and four planets within each group. I thought hard about one planet from each group being formed on just one line. Strong evidence of every planet existing in a spherical shape caused me to theorise that the lines must therefore be tubed or cylindrical or something similar to that. My theory of 'tubed magnetic lines' was about as much as I had achieved, until the urge for an answer began to eat away at my mind.

I stumbled on the answer after watching one particular internet video. I recall it today as, **'The Thin Wedge Which Opened the Door'**. I went to the internet (as we sometimes do when we need answers), and within just seconds, I found a video of a man who placed a very strong magnet into a glass of water and froze it overnight. (Sorry, but I can't find the video today.) The video was of a Scotsman who had been consuming a significant amount of Scotch, or at least he sounded like he had, but a Scotsman drinking Scotch from a glass was no real feat of discovery. It was the following which truly hit home.

The Scotsman dropped this strong magnet into a glass of water and froze it overnight. In the morning, it revealed frozen **tubes** of ice projecting out from each magnetic pole. And so the Scotch glass was now filled with the evidence I needed to confirm my theory that magnetic lines undeniably existed in circular tubes.

I went back to the original four-lined diagram. According to that video, this TUBE theory was now an unproven science that I myself found extremely hard to conceive, and so I watched the video over and over and over again. I remember the moment when I concluded that it was my mind, for some peculiar reason, that blocked me from applying it to *The Holy Grail of Science*, which had been unfolding over the weeks/months/years. I began to also theorise why my mind wouldn't allow this 'tube' fact and figured out the following. Have you got that hat, folks?

It was that childhood experiment with the magnet again where the iron filings displayed a set of very flat lines due to it being displayed on the surface of the paper. It had somehow burnt this vision of into my mind (and I'm sure into each of yours) that magnetic lines are very flat, but I can tell you that they are NOT FLAT and that they undeniably **exist in tubes**. I then began to contemplate quite strongly that each planet was somehow formed inside the frame of a tube. I'll show you the evidence of why each line exists as tubes, which will in turn explain how each planet was initially generated as merely a sphere of light.

The magnetic component of the light frequency slowly intensified over millions or billions of years to eventually exist with a strong enough magnetic field component to begin attracting each of the first particles which were initially smaller than atomic particles. The evidence that I mentioned is in 'the spherical shapes of planets'. I began to contemplate the natural structure of each planet, and I finally realised after a week of watching the magnetic tube video that each planet had a number of spherical cores.

For me, Earth was the greatest example of multiple cores; in fact, our Earth has the highest number of cores along with every other metallic planet. Therefore, if our Earth was formed inside a tubed frame and our Earth had multiple cores, then the tubes were also in multiple-tubed frames. The next step was to conclude that multiple concentric cores within our Earth meant multiple concentric tubes within magnetic lines, and our Earth (as I said) is the greatest example we know of with multiple concentric cores. The multiple concentric tube theory was not, however, as evident in the next planet, which was formed on the same line as our Earth and out from our Earth. It was a giant gas planet, and all gas planets are recorded by science as having just two cores.

So spherical multiple-concentric-cored metallic planets were formed by multiple concentric magnetic tubes, but how did it actually occur? Before I show you exactly how each planet formed within multiple concentric tube frames, let's just look at the strong evidence which supports the theory that the planets were at least formed on the actual lines, described in the following reveals.

This chapter describes the actual mechanism in which each planet was generated. I've just described the mechanical structure of magnetic lines existing in multiple concentric tubes, and I briefly explained that each planet (especially our Earth) was formed with multiple concentric cores. The following describes how our Earth (as the best example) was first generated as merely a light frequency which possessed an extremely miniscule amount of resonant magnetic energy.

In the chapters above about magnetic energy, we know that electrical energy is also always present, but I can tell you that the four lines existed primarily in an extremely high percentage of magnetic energy. I need to now repeat a fact mentioned earlier but with better detail as it is of utmost importance. Every electromagnetic light frequency on the EM spectrum is an electromagnetic light frequency of energy, travelling in waves made up of photon particles. See http://

www.physics4kids.com/files/light_emtype.html. Please read just the first paragraph.

Every gamma wave energy is a light frequency, including every other electromagnetic frequency like radio waves and everything in between. Every electromagnetic energy most probably travels at the speed of light but only under a vacuum, which is in space. The fact of photons being the actual physical properties of every frequency of electromagnetic energy is the KEY here.

Each of those six largest planets (that I mentioned in an earlier chapter) is recorded in the ascending line chart and reflects colours of the visible light spectrum, which are red for Mars, blue for Earth, yellow for Jupiter, orange for Saturn, indigo for Uranus, and purple for Neptune. As I've previously mentioned, they are just a few. Other light frequencies also exist on the EM spectrum, and they also created planetary points of light frequencies which grew into a stronger percentage of magnetic field energy, developing into planets. (More on those soon.)

Now, so that you will hopefully understand this better, I will refer to those six spectrum planets as a **visible spectrum of light frequencies**. I need to also gently remind you again that the visible light spectrum is in the centre of the EM spectrum. The list of frequencies on the EM spectrum are recorded in a line, and this line is actually the description of a full magnetic line or an unbroken line or an unfractured line. Each of the two original lines which became divided (now appearing as four) was once just single lines that crossed over, and each single line was originally a full line of frequencies on the EM spectrum. Later, I'll describe how it was possible that each planet (especially our Earth) began as merely a light frequency which was generated when a fracture occurred on one of the original two magnetic lines when it began to curve.

Each light frequency possessed an extremely miniscule amount of magnetic energy, which then strengthened during millions or billions of years to eventually become strong enough to start attracting larger particles. The following link below is to a video which lists the subatomic particles that make up an atom: https://youtu.be/k6Rclo5EQGM. My apologies, but the video is of very, very poor quality but does contain the information on the list of subatomic particles. This link here is the full video of the documentary, including the short video from the link above: https://www.sbs.com.au/ondemand/video/845536323654/inside-cern.

After the rotation velocity increased on the centre mass, each original line curved and then fractured in steps which were in the magnetic ratio and equal distances apart just like a rainbow or a visible light spectrum. Where each line fractured, it generated each planet as merely a light frequency, especially the six coloured planet point fractures listed above, which were generated out of the original line on the list of visible light frequencies. The planet fracture point opposite the frequency of our green Earth (on the original line) was the next colour as blue on the visible colour spectrum.

I can tell you that the metallic planet which was generated on the opposite side of our Earth (on that same paired Line) was Venus, but its recorded colour is somewhat ambiguous today. I believe the ambiguity in the recorded colour is because it is hard to see between our Earth and our sun. The intensity of light being emitted by our sun is therefore causing the ambiguous distinction of its recorded colour.

Here are some simple facts about all the planets growing inside magnetic-tubed lines. If all the planets were formed within the frames of multiple concentric tubes, then the natural design of this magnetic structure meant that each planet was (more or less) 'held' by the multiple concentric-tubed frame whilst each planet grew within each frame. *I identify the mechanism which held each planet in place on each line as* **magnetic possession**.

The best way that I can describe the magnetic frame structure of the lines which were multiple concentric-tubed frames is to suggest that their magnetic structure was very much like the physical appearance of a Russian stacking doll. (My apologies for my lack of ingenuity, but this description may help before a video is made.) See https://en.m.wikipedia.org/wiki/Matryoshka_doll. *I therefore identify the magnetic structure of these tubes as the* **magnetic matryoshka structure**.

I began to theorise or explore the cause of each metallic planet (especially our Earth) having multiple cores, but the next planet, the giant gas (formed next to each metallic on the same line), only had two cores according to science. By the time the planets were formed, the two lines had divided into four lines.

Our Earth was formed on its line (as the first planet on that line with three cores), and the next planet was a giant gas planet which had one less core; then logically, after our Earth was generated, a tube was therefore dissipated within the line which then generated only two cores at the very next planet as a giant gas planet. It is the same for every metallic planet with three cores and every giant gas planet with only two cores.

Again, a smaller number of multiple concentric-tubed frames existed after the first fracture occurred. One tube was therefore dissipated after generating each metallic planet, especially after our Earth was generated, which is the best example. I'll soon explain exactly how the magnetic resonant mechanism of a light frequency was generated, which I have described (until now in lay terms) as fractures. You will clearly see in the next chapter how the metallic planet fracture points caused the dissipation of a tube in a very simple but extremely logical description. I'll also show you soon how it was possible that the multiple concentric tubes generated each planet point as a light frequency possessing magnetic resonant energy.

So please take a small break again. Maybe have a recap of just a few of the chapters above and come back with a fresh mind to view this incredible reveal. I suggest the hat again, maybe even two hats (one for you and one for me) for the following. This is also probably the most technical reveal that I have to describe about the tubes, but it is really quite simple to understand, how it was possible that the multiple concentric tubes generated each light frequency planet point of energy.

Our sun's lines were once straight. After the centre mass increased in reactive energy, thus rotation, it was then followed by increased velocity. Each line began to suffer from that increased resistive force (on each front face), and they were forced to begin curving. We know from the cane that the curve began from the centre mass and then travelled (in linear form) out to the first planet point where a fracture occurred, and this first fracture was where each metallic planet was generated.

As these lines of multiple concentric tubes began to curve, the outside tube of each line began to curve first because it was the longest and of the largest circumference, suffering a greater resistive force first. Each outer tube was also facing into the wind (so to speak), and the other inner tubes were protected from the wind whilst being inside the outer tube. This outside tube began to tilt (due to its rigid mechanical structure) but then curve to a certain degree when it then touched the next inner tube. It eventually touched the next inner tube at two points of contact, which were on each side of the next inner tube; and where each concentric tube touched one another at these two points, there was a crossover of each tube's outer tube structure.

The result was an extremely minimal amount of energy generated as a light frequency which initially consisted of a circle of light energy that eventually grew in magnetic strength during millions or billions of years, eventually becoming strong enough to begin attracting particles to fully form as each planet. The first particles which were attracted were super-subatomic particles (smaller than those listed in the CERN video given in the last link), which also had their own magnetic field. These particles aligned with the extremely minute energy within the magnetic field that existed within the light frequency. Those super-subatomic particles plus larger, then larger particles and larger particles over time, all began bonding within the light frequency from the moment that they were attracted.

Both the light frequency which emits the blue energy of our Earth and the magnetic poles of our Earth grew in field strength as our planet's mass also grew in particle size and field strength. The poles of each planet (including our Earth) grew in strength and mass as each accumulated magnetized particles and grew exponentially in mass, with each planet growing over millions of years to eventually become what it is today. You see, the accumulated particles which began to be attracted had their own frequency and polarity.

I can tell you that, on our Earth, each particle aligned with each polarity of our Earth and slowly grew over millions of years as the North and South Poles of the planet. Most importantly, particles of a neutral polarity were also attracted and accumulated into the twelve-inch neutral band of energy which is positioned at our equator. This is a very wild theory that I have about the neutral subatomic particles, like that of the neutrons of an atom. It's a theory which states that neutrons do not have a neutral energy and that it is an energy which is just as substantial as a proton (+) or electron (-).

As you all know, when white light is projected through a glass prism, a refracted spectrum of light in seven colours is projected out of the other side of the prism (a rainbow). *When an outer concentric magnetic tube curves over and touches the next inner tube, it produces what I identify as the 'refractured' magnetic tube.* It is **NOT**

REFRACTING as light does; it is a refracturing magnetic tube under my new discovery. I identify this mechanism as refracturing because magnetic lines are not an identical in structure to white light, although the two are definitely directly related. Refraction of white light does possess magnetic energy, but it is extremely low.

I can tell you that light energy and magnetic energy share the same relationship as magnetic and electrical energy, which is identified as electromagnetic energy. When the light energy outweighs the magnetic energy, it is identified as light; and when the magnetic energy outweighs the light energy, it is identified as magnetic, but the overall energy does contain both energies. Magnetic tubes **refracture,** under my new theory about magnetic line energy, along their length in steps which are in equal distance according to their magnetic ratio distance. A rainbow's colours also exist in equal distances between the colours in accordance to the identical magnetic ratio distance. The equal distancing and the magnetic ratio distance are common facts between the two separately functioning mechanisms.

Again, our sun's lines are very long, and they do possess a line of colour frequencies in them (the EMS), which are made up of both the visible (seven only) and the invisible light frequencies. The energy which existed within our Earth's generated refractured planet point was not as powerful as the energy generated by our sun's magnetic pump within the crossover, but the planet point did magnetically function under much the same principle as magnetic resonance. Our Earth's refractured field where two points of magnetic resonance were generated was not as powerful as our sun's magnetic pump crossover largely because light frequencies only possess an extremely low or ultramicroscopic level of the magnetic component within each electromagnetic frequency. (Remember the scales.)

In closing, magnetic energy (which is just one component of the two energies within electromagnetic energy) very obviously does possess every colour of every light frequency within the energy. So the lines which radiate from our sun began as just two lines and finally functioned magnetically as four lines after reactivity intensified after millions or billions of years. Each of the diagram lines are primarily magnetic energy (with an extremely small amount of the electrical component), and let me repeat it—magnetic energy does, in fact, possess every colour in light frequencies.

I'm going to take the example of just the seven visible light frequencies positioned in the centre of the EM spectrum which once existed within each of the original two crossed-over magnetic sun lines.

Those two lines once possessed or contained every visible light frequency in the centre area of each electromagnetic line of frequencies, and the centre band was where the first eight largest planets were generated after the crossover expanded. Quite simply, each of the original two lines were just two full electromagnetic spectrums which crossed over, melded together in the centre as one, and began to curve and refracture at points along each line which were in steps or stages under the factor of magnetic ratio distance. These steps or stages where planetary points were generated were at equal distances just like a visible light spectrum or rainbow.

The physics of refracturing generated magnetic resonance and began accumulating particles or materials. So as to help the reader understand the following theories, I will describe the following basic functions of an 'electrical thermocouple', which is an electronic temperature field sensing device that will transmit incredibly low signals to a digital display. **The mechanics of a thermocouple are these.**

Two wires are twisted together at the ends. One wire is made of copper and the other of nickel, and a variation in temperature will generate a voltage differential at the junction which is then amplified and sent back at a digital display. In brief, each metal will generate a different potential or voltage at the very same temperature. In essence, a thermocouple measures voltage differential, which is dependent on temperature.

The device functions very well because every metal can generate a varying voltage when subjected to varying temperatures. Copper and nickel are used because they corrode the least, and the temperature differential is best suited for the application. The two dissimilar metals twisted together at the junction will each vary in voltage when exposed to the ambient environment. It's quite incredible but has been around in electronics for years.

What is important here in helping understand the next theory is the manner or mechanism in which the actual electromagnetic force (EMF) or voltage differential is generated by the simple junction of two dissimilar metals. The junction will produce a voltage differential between the two metals. Each metal will produce a different voltage due to the temperature it is subjected to.

The junction of the two metals can emit a voltage differential anywhere between 0.025 volts and 0.035 volts, which is in the millivolt range. One millivolt is equal to 0.001 volts. Keep in mind the actual manner or mechanism in which this miniscule voltage is generated whilst we look at this next reveal.

The following describes how each planet was first generated from merely a light frequency when the lines refractured. Each planet began as merely a light frequency which possessed extremely low magnetic energy. Each planet point eventually grew to a strength where it was then possible to begin attracting super-subatomic particles and then subatomic particles (seen in the video link from CERN), atoms, molecules, extremely small dust particles, ultra-micro-sized dust particles, micro-sized dust particles, larger dust particles, larger particles, sand-sized particles, small rocks, bigger rocks, large meteors, and so on. (I'm sure that you understand how they grew in size.)

Each one of these attracted particles existed primarily in one of three polarities—north, south, or neutral. As previously mentioned, the polarities of attracted super-subatomic particles which make up our Earth consisted in negatively charged particles and positively charged particles in exactly the same manner that an atom consists in negative electrons and positive protons; plus, there were also neutrally charged particles which accumulated at the equator of our Earth (in the twelve-inch band).

Somebody recently mentioned to me that an area of land that they knew of in South Australia was extremely magnetic, and their description did not resemble a ley line, which is sometimes called a meridian line. I believe that because this area was so large that it was actually where a giant extremely magnetic meteor had landed during the extremely soft period (the newborn or early Earth stage). The giant meteor had melded into that soft landscape when our Earth was still accumulating as Gondwana or just after. I refer to our Earth (after Gondwana, which was pool-balled and then broke up into many continents) as 'early Earth', which was best described as 'newborn Earth' as it was very soft, having just come off the paternal line.

Earth started to become a planet, and it eventually had a structure but was still growing as a mere light frequency. The magnetic resonance eventually, over millions or billions of years, became strong enough to start attracting atoms (due to magnetic resonance) which were primarily carbon atoms. Carbon is said to be an atom which is capable of a quadruple bond, meaning it can bond with three other atoms to eventually become a molecule.

As previously mentioned, all the light frequencies exist in a band or range. The reason why carbon is our primary atom is that, when we look at the frequency of blue, it very obviously exists in a number of bands, and carbon is an atom which is the most resonant within the range or bands of blue. Carbon is perfectly in tune with blue. Our Earth originally existed in the range of frequencies of blue light, and thus, atoms that resonated within the bands of blue were attracted. The particles which were eventually attracted to form as our Earth began to increase in size and mass to eventually become that one land mass identified by science as Gondwana.

Let me say again now that our Earth exists in a frequency of blue light, and it does emit the colour or light of blue energy. All the particles/materials in the crust are in the strongest frequency of blue, and if you have ever seen iron that has been cut by a saw, you might notice that the face of the cut has a bright blue tinge which is actually quite colourful (just like our planet's blue hue seen from space). Also, when steel is being welded, it emits a bright blue flash; and when welding aluminium, the welding flash is very close to white. And steel is often referred to as blue steel or blue metal.

All the other frequencies (along the length of each refractured tubed line) were also in resonance with the attracted materials to eventually become those planets and were also a direct result of magnetic resonance. The light frequency of magnetic resonance at Mars is red. Mars also began attracting atoms, molecules, and bigger particles which were also in that identical frequency to finally become Mars as we know it today. Jupiter did the same, along with Saturn, Uranus, and Neptune; but Uranus has a very interesting story. It is in the colour frequency of indigo or light blue, which is close to being amethyst.

Science recently announced the discovery of a hexagon-shaped moon that orbits a planet in our solar system. And now as I glance across the room where I sit, I can see a very large amethyst crystal that is actually very much an indigo or light blue. The planet Uranus was formed by a massive amount of crystal compounds, and when it was punched off the paternal line by another planet, a large chunk of Uranus shattered off and became that hexagon moon.

I have mentioned in the early part of my book that each planet has its own pulse of energy. Our Earth has a pulse of 8–12 Hz, and this is generated by the two magnetic poles which exist deep in the molten centre core. These two poles were created when the materials were attracted and settled into our planet's mass in their perspective polarities. The crust of our Earth emits the frequency of blue and is seen as the light source surrounding our planet. (I am saying that our sky is the colour blue due to this but more on that in a moment.)

This blue energy is due to the circle of light energy which grew in its component of magnetic resonance that was generated inside the outer tube. The frequency of the colour blue is generated and emitted by the outer core (by the crust) because it was the outer tube that created the circle of that frequency, which then strengthened. Our Earth's hue of blue energy **does** actually extend out into space as seen in NASA photographs (https://goo.gl/images/2LgTpM). Again, if the link has gone, just search for the blue hue surrounding our Earth.

I can also tell you that, according to science, our Earth's atmosphere exists in six different layers, beginning with the troposphere through to the exosphere, and I say this is due to all the layers being in differing bandwidths of the frequency of blue. Each band of blue is a reflection of the blue light frequency, projected against each atmosphere's photon particles. You can clearly see in those photographs by NASA that the blue frequency is quite a strong light energy at the outer hemisphere but is just ever so slightly a darker shade than our Earth's sky. This link clearly displays other shade variations of the blue hue as it projects the varying emitted energy out and into space: https://en.m.wikipedia.org/wiki/Outer space.

The actual frequency band of blue emitted by our Earth does, in fact, vary, and that's because every shade of blue exists throughout the range of the frequency of blue. I can tell you that the sky is that shade of light blue because the frequency band of the colour blue is being emitted from the crust, projecting outwards into space, and it is light blue in that region where our sky is positioned. Again, each minute dust particle in each layer of our Earth's atmosphere has the frequency blue projected onto them, which is emitted by the blue hue. *The frequency range of the colour blue varies from 606 to 668 trillion Hz, and this is made up of every variant of the frequency, thus the shade of blue.*

Our Earth is made up of as many different atoms which can each form bonds with other atoms. I'm going to go out on a limb here and say that maybe every metallic planet is made up of atoms that can also make quadruple bonds, but I'm not totally sure of that one. I think that due to each metallic planet existing in a band of frequencies, which exist in the same range of atoms, they can make quadruple bonds, although some light frequencies exist in a range that has a larger bandwidth than others, and maybe these frequencies which became planets are of a greater number of accumulated atoms and molecules. Green has the widest bandwidth, but I am only theorising this right now.

When each line began to curve, the outer concentric tube touched and formed a crossover with the next inner concentric tube. The crossover occurred at two points where the tubes made contact, and light frequency point with an extremely small amount of magnetic resonance was generated. There were two points of contact, one on each side of the tubes, and they were therefore in two places, which then formed a circle in that particular frequency that possessed very little magnetic resonant energy.

The level of energy had to grow and was generated in much the same way as our sun's energy with one difference, which is quite significant. Our sun emits white light from the crossover of two full magnetic lines. Where the two sun lines cross over, these two lines are unbroken and therefore emit an accumulated or combined range of frequencies which together make up white light in exactly the same way that seven colours make up white light. The recorded colours of all the other planets (besides these six in a row) are a little bit ambiguous today, but I can tell you that ALL the other planets, including the seven micro-dwarfs, will all logically exist in each one's own frequency of colour.

Facts about the Frequency of Every Magnetic Line That Exists

Absolutely every magnetic line projecting out of absolutely every pole does exist in the identical frequency which is made up every frequency of electromagnetic energy. Each magnetic line is an accumulated or combined number of frequencies consisting in every frequency of electromagnetic energy.

Let me show you the first evidence. The basis of this is the actual scientific description of the electromagnetic spectrum, where every frequency on the EM spectrum is a light frequency, from gamma waves to radio waves (https://goo.gl/images/eVasmg).

The second supporting evidence is that every sun was generated from just two full lines. They are now crossed over and are attracting hydrogen atoms under the laws of magnetic resonance. The proper reference actually is electromagnetic resonance. The principal foundation of resonance is that each of the frequencies must exist in the absolute identical frequency.

The third is in the example of our sun, which was also generated from two full lines which refractured, generating every planet within each light frequency; therefore, each line is an accumulation or combination of every frequency of electromagnetic energies.

The fourth and most prominent evidence is that every magnetic line consists in identical combined frequencies (white light is an accumulated frequency band). The planet alignment every 500 years is also a frequency alignment where every planet's frequency is synchronised (when the planets link up whilst passing one another). This is to point out that extreme weather events become even more extreme after a frequency alignment, meaning that a pure resonant frequency exists after the alignment.

The suggestion on the reported weather events shows that the magnetic energy of every planet existing in the identical frequency is extremely strong. Every hydrogen atom exists in the same frequency in resonance with white light. The evidence of that is seen by the white light which is emitted by the ignition of our sun's attracted atoms of gas.

Again, each of the two original lines which crossed over and generated our sun as a magnetic pump (before curving and refracturing) was a full line of electromagnetic energy, totalling the full amount of light frequencies on the electromagnetic spectrum. When the two original magnetic lines eventually began functioning electromagnetically as four lines, they steadily began segregating within the centre mass, functioning as four lines, and then expanding, curving and refracturing. When each full line separated, they segregated into separate frequencies which emitted the colours and properly formed each planet.

The above was the event (I mentioned in previous chapter) briefly described as the expansion of each of the two full lines when curving and then refracturing occurred, with pairs of planets being generated on what WAS once a full line but now separated into frequencies. Each full refractured line created one full line of planets, and there were two full lines. And there are now therefore two sets of planets originally generated that still exist today. The total number of planets was originally made up of two lines of ascending planets from two full spectrums of visible and invisible colour frequencies.

THE LIST OF SIDE EFFECTS CAUSED BY THE LINES

These chapters list the side effects caused by the lines when injecting or projecting into our Earth, revealed in an order that is easiest to digest. *The list is very extensive and real due to their extremely logical description, so please take them slowly.*

I need to again describe some theories about the planet alignment with further facts added. **The planet alignment was a mechanism which also generated the alignment of their frequencies** due to all the planets being aligned in their magnetic energy. The realignment of their frequencies is the primary cause of the extreme weather events on our Earth ever since the planetary alignment. Let me try my best to explain the whole scenario.

The bottom line here is that extreme weather events have intensified ever since the planet alignment occurred. Tornadoes have become more intense, sinkholes have occurred more often, and storms have occurred with greater fury than before the alignment. First, I will describe the extreme weather events which have occurred lately, but I also need to point out that the extreme weather occurring before the planet alignment has increased in force after the alignment.

Wouldn't you agree right now without too much hesitation that the weather has intensified on our Earth since 2000 but especially during 2015–2018 after the alignment compared with the weather events which occurred in 1995–1992, which was the same amount of time before the alignment? When I mention 2015–2018, I am talking about fifteen to eighteen years after our alignment. When I mention 1995–1992, I am talking about the same amount of years before our alignment. You only need to check the online statistics as far as increased weather is concerned. The finer details are in the chapters on extreme side effects of the lines which cause weather and are coming soon to you in very descriptive stages of reveals to clearly show you why.

The greatest examples of incredibly extreme weather events or those which have increased in force were briefly described in the previous chapter about eddy currents on our Earth as tornadoes. Tornadoes have been incredibly intense after the alignment. I also previously described recent sinkholes which have increased in intensity. They have notably and incredibly intensified during 2015–2018 after the alignment. Massive floods are another intensified event, as well as volcanic eruptions, which I will explain later. You will be totally amazed at the evidence I will provide surrounding all these intense events in a later chapter.

Some of you may be asking yourself right now, 'How is it possible that every planet's energy and frequency causes weather, or how could the planets even be out of synchronisation in the first place?' So as to assimilate each planet as a source of magnetic energy which has a small supply of electrical potential, I must first tell you that, in the electrical supply industry, when paralleling two separate sources of electricity, we must use a special electrical meter which displays each electrical source's sine wave and then align the two sine waves before closing the switch.

One supply of electricity might be the main source coming from the power station. So as to boost the supply of power for that area of the city, they might then bring a power generator online, but the sine wave of the generator must be in perfect alignment with the power station supply before the two sources can be joined up. By the term 'joined up', I am saying that the generator can be used on the power grid supplied by the power station.

I need to describe the full details of exactly what occurs during a planet alignment in slow reveals later in the book before I can disclose how weather can intensify, but believe me, the evidence supporting a frequency alignment occurring at the same time is quite overwhelming and real. It's a very hard description for me to put forward in words, but it all began when I noticed that there was quite a change in the weather events which have increased in force ever since the planet alignment.

It is all based on another fact that I touched on briefly. Our Earth has a pulse of just 8–12 Hz that is emitted by the north and south magnetic polarities which also have a positive and negative electrical charge within the centre core, but the blue colour is emitted by the source of the outer core. I mentioned in an earlier chapter that two different sources of electricity which exist in different frequencies can be transmitted within the same cable. The transmission of two totally different electrical sources (down the same cable) occurs several times a day in the grid system here in Australia, when another signal at 1,050 Hz is transmitted over the top of the normal 50 Hz supply of the grid. This other supply of 1,050 Hz transmitted down the same cable communicates with the street lights and hot water systems throughout the grid system.

I hope you can clearly see that it's possible for two different frequencies of supply to exist at every planet, especially here on our Earth. One supply is the centre core with a pulse, and the other is the supply of light energy being emitted by each planet. Let's take a brief look now at the energy field of our planets and our sun.

Our Sun's Magnetic Energy

Our sun was recorded by early science as having four magnetic poles (displayed in the diagram sketch) which project out from the outer layer or core of our sun. In accordance with the common laws of magnetics, every magnetic pole has three strongest open-circuit lines projecting out and then stopping at a distance away, and the magnetic lines in the diagram sketch have a large amount of missing lines, including every other secondary open-circuit line.

Remember that when we consider the orbital path of our Earth, we now know that there will be three strongest open-circuit lines which our Earth will pass through, but there are, of course, other open-circuit lines projecting from each pole face. There are also a greater number of closed-circuit lines (under common magnetic laws) that project out from the South Pole and return to the opposite pole, dissipating the energy.

Our Earth's Magnetic Energy

As previously mentioned, our Earth has two magnetic poles (North and South), and each possesses the stronger open-circuit magnetic lines which project out from each pole and then simply stop at a predetermined natural distance away. However, each of our Earth's ley lines (which are the closed-circuit lines) travels throughout our Earth, especially the outer crust, from one pole to the other and exists as the weaker dissipated energy. The molten centre core of magma of our Earth is recorded by science as the source of our Earth's magnetic field, although I clearly remember from school that a heated magnet is not as strong in magnetic energy as the same ambient magnet.

Just briefly, if the nucleus of an atom was the size of a basketball, the electrons would be thirty-two kilometres away. Realise that all that empty space is taken up by the energy field of the atom in a very similar way to our sun and its family of planets. The energy field of each atom and the combined energy fields of every molecule in our Earth's outer crust are the source of its magnetic energy field which exists within the materials.

The next layer within our Earth is the mantle, and it too has a magnetic field, with open-circuit lines projecting from each pole and the closed-circuit ley lines travelling from one pole to the other. The energy of each particle (of the matter) that the mantle layer consists in also constitutes the magnetic energy field of the mantle.

The next inner layer of our Earth, which is the molten centre core, also has a common magnetic field, with open-circuit lines projecting from each pole and closed-circuit lines running from pole to pole. This molten centre core is recorded by science as the source of our magnetic field and as the source of the pulse of around 8–12 Hz. The centre core also has closed-circuit ley lines of magnetic energy projecting out from the South Pole, travelling throughout the centre core and the next layer of the mantle, and returning to the North Pole. The magnetic field (consisting in the magnetic energy of each particle of the matter) that the centre core consists in is also its magnetic field.

Our Moon's Magnetic Energy

Our moon also has two magnetic poles which have open-circuit lines projecting out from each pole and then also stopping at a predetermined distance away (just like any common magnet). Our moon also has closed-circuit lines that travel from its south pole throughout its materials, especially its outer layer, and then return to the opposite pole. Our moon consists in one spherical complete mass without any cores, and it also has a magnetic field consisting in each particle of the matter that our moon's mass is made of, having once been a chunk of Mars which broke off.

We now know that our moon has a gravity field which is about 17% of the strength of our Earth. This weaker field is very obviously due to Mars not having a very strong field and due to the laws governing the breakaway by chunks of magnets. If a chunk breaks off a magnet, the chunk is somewhat weaker.

Magnetic Energy Compared with the Gravitational Energy of Our Sun, Our Earth, and Our Moon

In my previous chapter on king and neap tides, I stated that our sun and moon generate each daily tide. I also mentioned that science describes king and neap tides as being generated by our sun and moon, but contrary to this, I said that the three strongest lines as the source of those three strongest tides.

Our sun's gravitational energy is recorded by science as the source that holds our family of planets in their orbit, and each planet's orbital pattern is described as 'slightly wavering'. Our Earth's gravitational energy is recorded by science as the source that holds our moon in orbit around our Earth, described as a 'slightly wavering elliptical pattern'. The elliptical pattern of our moon is a path (more or less) in the shape of an egg with our Earth at one end. Our moon's gravity is recorded at 17% of our Earth's gravity.

The source of gravity is the closed-circuit energy lines of our Earth which is a great deal weaker than the open-circuit line energy. But exactly how the closed-circuit lines generate the gravity field is a subject for another day. I assume that it has something to do with my earlier theory that closed-circuit magnetic line energy has a higher percentage of electrical energy within it because the closed-circuit lines have a far greater flow. Another theory I have on gravity is that our Earth has a number of cores with their own magnetic fields which overlap with one another, and this may be the mechanism which constitutes gravity.

In a previous chapter, I mentioned that science also summarised the theory that our moon was once somehow a part of Mars after a chunk of Mars broke off. I offered my theory about Mars, suggesting that it collided with our Earth and that our moon was a chunk of Mars which broke off. Our moon's gravity is at 17% of our Earth because it is just that chunk of Mars, and our moon consists of just one solid mass (more or less) like a giant magnetic rock without any other cores, plus two magnetic poles with very weak open-circuit lines projecting out from them and stopping at a predetermined distance away. Our moon also has closed-circuit lines (just as every common magnet has) travelling through its outer layer from pole to pole as its ley lines. It is just the inner part of our moon which was that chunk of Mars that broke off, and the outer layer has accumulated over billions of years to become the sphere that we see today. It also has a magnetic field in the matter or atoms of the materials of which the outer layer of our moon consists, also having a low magnetic charge. The gravity or magnetic field of Mars is low in comparison to our Earth.

You may remember from those simple experiments in school that when you break a chunk off a magnet, a new magnet is formed by the chunk, which has lower magnetic energy. In simple terms, the new chunk of magnet always has a lower charge of magnetic energy. Also, remember from school that to strengthen a weak magnet (produced by a chunk), you must then stroke it with a much stronger magnet so as to induce it with a stronger magnetic field.

The chunk of Mars, as our moon (like any other chunk of a magnet that breaks off), has a very weak magnetic charge compared with Mars, having never been induced by another stronger magnet. The low energy of our moon may also be due to the type of matter or atoms of the materials that were attracted to create Mars as a planet. The matter or atoms of the materials which Mars is made up of may also have low magnetic energy. The materials which were attracted to finally make up each planet are very different for each planet. Our Earth in comparison to our moon has a greater number of cores or layers which provide an increased source of our Earth's magnetic field, and I believe that due to this, our Earth has a far greater gravitational force.

Again, I agree with science and say that our Earth's 'daily tides' are a result of an influence by the energy of our sun and moon after inducing our oceanic waters. But science does not specify whether it is gravitational or magnetic energy that influences and generates our daily tides. Our Earth's daily tides are actually generated by the combined gravitational energies of our sun and moon, which is the source, and not just merely magnetic energy. Also, I reiterate that the three strongest tides every three months are a direct result of the three strongest magnetic lines inducing our Moon which in turn induces our Earth's oceanic waters.

Allow me to show you a little evidence of the combined gravitational forces which generate daily tides. This clearly shows that it is not magnetic energy which generates daily tides. When our Earth's daily tides are generated by our sun, our Earth is not close to any of the energy lines (seen in the diagram) which would be radiating from the outer surface of our sun as the open-circuit lines. Therefore, if our sun's open-circuit magnetic energy is the source generating daily tides, then our Earth should be closer to our sun.

In regard to daily tides generated by our moon, remember that our moon has open-circuit lines which project out from each pole.

As our moon orbits past our Earth (in that elliptical pattern), at those two moments when the moon's open-circuit lines (projecting out from each pole) are most likely pointing at our Earth, there is only a relatively small window of time when it is possible for our moon's open-circuit lines to possibly generate a daily tide. All this is strong evidence that it is our sun's and moon's gravitational energy which generates each daily tide.

From the list given above (under the three headings) identifying the various sources of magnetic energy, I believe that the actual source of energy that generates each daily tide is gravitational, notwithstanding each of the three strongest tides every quarter year. From the four lines of the diagram, my theory arose that each line is an open-circuit line projecting from a magnetic pole, and there will, of course, be a greater number of weaker closed-circuit lines which radiate from pole to pole that are missing. Remember, earlier, I described <u>closed-circuit line energy</u> as existing in a greater percentage of AC electrical energy which possesses a frequency. As previously mentioned, closed-circuit magnetic energy lines have the highest rate of magnetic particle movement of any magnetic lines, which in turn causes a higher degree of movement of electrons.

Our sun's gravitational energy, which holds each planet in orbit, has a direct relationship to the frequency of each planet; after all, an orbital period is the time it takes for a planet to complete one cycle. My theory here is that a planet has an orbit frequency due to a resultant force which is gravitational which has a high degree of energy that possesses a frequency. As we can see, our sun has four lines, which are open-circuit lines of the diagram, and it will also have other open-circuit and closed-circuit lines outside of these, which will be spread throughout our solar system. Of course, a strong indication is that the closed-circuit line energy of our sun is generating many different planetary orbits due to this energy somehow possessing many different frequencies. My new theory that closed-circuit lines possess a greater percentage of electrical energy is a very good similar evidence of <u>gravity being a combined magnetic energy</u>.

When referencing closed-circuit lines possessing a frequency, I need you to visualise a diagram of a magnetic field. I'm sure you can recall a bar magnet with the open-circuit lines projecting out of the pole ends and the closed-circuit lines travelling from pole to pole but with a significant curve to the other end of the magnet. The image in your mind doesn't require any great accuracy. Just the simple visualisation of those open-circuit and closed-circuit lines is sufficient.

To go one step further, you need to visualise just one closed-circuit line and realise that they are all C shapes attached to the side of the magnet. Also, there are a number of C-shaped lines which are best described as concentric (each inside of the next). The identical number of C-shaped lines exist on the other side of the magnet, and they are also perfectly identical in a mirror image on the other side of the magnet.

Now take the centre open-circuit line (in your mind) and realise that these four lines of the diagram are open-circuit lines. Apply the image in your mind of C-shaped lines somehow into the diagram. Now visualise those C-shaped closed-circuit lines travelling from the surface of our sun (where one centre line leaves the spherical mass) and flowing to the next centre line of the diagram. With four lines radiating out from our sun and many closed-circuit lines in between, our solar system very obviously has an area where the planets orbit within a great deal of closed-circuit line energy.

Remember my theory that closed-circuit lines have a high degree of AC electrical energy and then look to the orbit period or frequency of each planet. If the closed-circuit lines have a higher degree of AC energy, then it would also be highly likely that the closed-circuit lines closest to the body of the magnet have the highest degree of

AC energy. What I am trying my very best to say is that the closer a planet is to our sun, the higher the frequency which will exist within the closed-circuit energy lines; and in fact, the closer the planet is to our sun, the faster it orbits around our sun as seen in the example of the closest planet, Mercury, and the farthest dwarf planet, Pluto. The closest planet, Mercury, has an orbital period or frequency of just 88 days for one complete cycle around our sun; and if we go to Pluto (the furthest planet), we will see that it has an orbital period of 248 years for just one cycle according to science.

I can tell you that each C-shaped closed-circuit line throughout our solar system which has a high degree of the electrical component (that exists in a frequency) is the cause of each orbit period for each planet. Each planet orbits along the identical path of a closed-circuit line. So by studying the planets orbiting our sun, we may be able to record the exact number of closed-circuit lines of every magnet. As a child, I can clearly remember the iron filings on that piece of paper (with the magnet underneath) with very jagged lines on the paper.

Science records the orbital pattern of our Earth and all the other planets in a very meandering pattern. This is caused by every planet following the jagged pattern of a closed-circuit magnetic line as it orbits. My theory here above is to tell you that, for some time now, I've considered the fact that the closed-circuit lines within our solar system possess a very jagged pattern, reflected by the orbital path of all planets and their moons.

Open-circuit lines which project from each pole are of the strongest flux or magnetic energy, and just those three centre lines alone have around 75% of the total energy with other open-circuit lines on each side of them, and so the total of open-circuit line energy may, in fact, be around 90–92%. The very weak closed-circuit lines make up the remainder as the other 8–10%, and as I've just mentioned, this is as a high degree of AC electrical energy. The supporting evidence is the orbit period of each planet, reducing as the closed-circuit lines reach further out into our solar system.

Every pole of a magnet also has an electrical charge which is positive at one pole and negative at the other; plus, the closed-circuit lines have a high degree of the AC electrical component. If we were to place two bar magnets side by side, meeting on opposite poles, they will attract each other due to the very strong energy of the open-circuit lines. Remember that the closed-circuit lines are an AC energy with a very weak pulse, and the open-circuit lines are therefore a very high degree of DC energy with very little AC and virtually no pulse.

I mentioned previously that gravity is a combined magnetic energy which exists in a high degree of AC energy due to gravity being generated by closed-circuit ley lines which exist in every magnet and therefore are the ley lines of every planet. Alternating current (which exists in a frequency) has a pulse and helps hold each planet in orbit due to the magnetic component of the EM energy which is in the closed-circuit lines.

Let's now look at our moon's orbital path. Amongst this theory above is the conclusion that our sun puts out an AC pulse which I identify as a combined magnetic gravitational force. Our moon orbits around our Earth due to the energy of our Earth holding it in that orbit. I described this energy emitted by our Earth as a combination of magnetic and gravitational energy, mainly caused by the influence of closed-circuit lines. Our moon orbits in an elliptical pattern, and at some point, our moon is at a distance farthest from our Earth. This is due to the combined magnetic and gravitational energy of our sun influencing our moon by pulling and pushing our moon.

As our Earth and moon orbit around our sun in different quadrants, it will be in changing polarities with each line in clockwise rotation as north–south, south–north. It is the rotating action of the diagram lines which began each planet's direction of orbit.

The Side Effects Of Our Sun's Magnetic Lines And Neighbouring Sun Lines

Eddy Currents

The following description is the effects of all the lines which cause tornadoes, hurricanes, typhoons, and cyclones. Here is a more descriptive reveal with regard to the information on how the side effects of sun lines cause eddy currents occurring on our Earth.

Tornadoes are always an atmospheric vortex with a tight circumference and often referred to as a twister in the United States. The tornado, hurricane, and typhoon which are always reported in the northern hemisphere do definitely spin in the opposite direction to a cyclone in the southern hemisphere. I've also noticed of late the occasional reference to a tight vortex occurring in the southern hemisphere as a tornado. Technically, whether it is a typhoon, a hurricane, or a cyclone depends on which ocean it occurs, and a tornado is a land bearing atmospheric vortex which spans from the cloud to the ground.

All the technical terms in the last few decades have become a tiny bit ambiguous, but I'll stick with the terms referred to in my opening paragraphs. Tornadoes, typhoons and hurricanes occur in the northern hemisphere and cyclones in the south. Each hemisphere generates a different direction of rotation but not due to the Coriolis effect. It is due to the direction of the eddy currents which also differs in each polarity or hemisphere.

Our Earth has North and South Poles, but magnetic poles also possess electrical polarities, positive and negative, respectively. The four lines of the diagram, plus the sun lines being projected from neighbouring suns, also exist in north and south magnetic polarities or positive and negative electrical polarities. The neighbouring sun lines are strongly influencing our planet, especially during the last few years and especially through the months of August to February 2018.

Every technical electrical person can tell you that eddy currents are a massive nuisance side effect which occurs in every electrical industry. Eddy currents occur in every industry from power station generator windings all the way through the distribution network and also in the tiny circuitry boards of electronics. The energy on a tiny printed circuit board is almost non-existent, but eddy currents are still ever present. Eddy currents will occur in just a cable that transmits electricity everywhere, from the power station's high-voltage cabling to high-rise cabling for buildings, as well as in printed circuit boards.

Very little is known about eddy currents in the electrical industry and how they are caused or generated. Eddy currents occurring within the electrical industry are also a vortex that is generated by the energy field within the transmission and are totally unpredictable. Our best chance to study the cause of eddy currents may be from our four sun lines (in the diagram) on our Earth. When the eddy currents occur, it appears that the stronger the magnetic field energy, the stronger the eddy current vortex that can and will occur. This is especially obvious when considering the four lines of the diagram or neighbouring sun lines that are projected onto our Earth.

In Australia, we usually expect the cyclone season to start around November, and that is the start of summer through to February, which is the end of summer, but it isn't the hot weather which creates a cyclone. My theory here is that the magnetic lines of our sun cause tornadoes through the events of eddy currents, and the neighbouring sun lines also cause eddy currents on our Earth as typhoons, hurricanes, and cyclones. The sun lines from our sun have a tighter circumference which cause the atmospheric vortexes of tornadoes that have tighter circumferences. The sun lines of our neighbouring suns which are much larger (look at the neighbouring sun which is 1,000 times larger) cause the much larger atmospheric vortexes of typhoons or hurricanes in the northern hemisphere and cyclones in the south, which are also large.

Typhoons, hurricanes, and cyclones are generated as eddy currents caused by the neighbouring sun lines. The neighbouring sun lines only point in our direction every twelve months which is when these large atmospheric vortexes occur. These large vortexes track the coastline during the night when the neighbouring sun lines are pointing onto our Earth. They continue to spin during the day under the energy of momentum.

Tornadoes are generated by our sun's lines; they only occur during the day, anywhere from just a few minutes to a few hours, and have a tighter vortex circumference. As I have stated, tornadoes occur all too often around the same time as king and neap tides but also outside those times. Please note that you will never see a tornado during the period between king tides. There aren't any energy lines around which are sufficient to generate a tornado in between king tides.

I am adamant to say that the four magnetic lines of the diagram cause tornadoes which are mostly recorded during king tide and neap tide periods and also by the other open-circuit lines on each side of those strongest tide lines but mostly around the time of king tides. Remember from the earlier chapter that the neighbouring sun lines are 1,000 times bigger than our sun, but more importantly, our solar system is rolling around the inside of a group of four other (much larger) solar systems. This is an indicator that our solar system would be near just one other solar system rather than in between two neighbouring solar systems.

I would like to make a safe bet here and say that 250 years before and after a planet alignment is when we are in between two solar systems. According to my theory of eddy currents caused by solar system magnetic lines being projected onto our planet, the fact that it is summer is not just the cause of cyclones in Australia or Africa, where cyclones also occur every twelve months. The same can be said for hurricanes and typhoons. Remember that all our solar system and beyond is one monstrous clock. Imagine the sun line from a neighbouring sun being projected onto our Earth. The reoccurring timing is important, and the timing of a neighbouring sun line means they will project into the same relative area every time.

A high percentage of cyclones occur on the north-east coast of Australia as well as the west coast but also on the north-east coast of Africa but not as often. Both of these locations are on the same longitudinal axis. They are both in a direct east–west region of each other. Whether it is a neighbouring large line (which causes cyclones, typhoons and hurricanes) or a smaller line from our sun (which causes tornadoes), this determines the size of the vortex.

You may have noticed that a cyclone (which can last for up to a week due to the momentum) is often seen tracking the coastline. Remember that the neighbouring sun line is only projected onto our Earth at night when we are facing the neighbouring sun (in the northern hemisphere of our solar system) here in Australia or Africa. I have noticed that a cyclone will only track the coastline at night when the neighbouring sun line is pointing onto our Earth and locking into the cyclone. During the day, the cyclone may well have drifted or meandered out to sea or any other direction just slightly, but it will always come back to the coastline or thereabouts during the night.

The neighbouring sun line tracks the coastline because these neighbouring sun lines once carved the continents into their final patterns. This occurred when the neighbouring sun lines were much stronger in energy. After millions of years, the neighbouring lines have finally moved their projections off the coastline and now follow the outline of the coastline but slightly out to sea. I will tell you much more about HOW these neighbouring lines carved out the final patterns of the continents in a later chapter.

The direction of rotation is a determining factor by the polarity of that particular line being projected onto our Earth, plus the polarity of our Earth which is facing the magnetic line at the time which will determine HOW the line is reactive. The extended evidence is overwhelming. All these vortexes reoccur in relatively the same regions, and each of the magnetic lines would be pointing into relatively the same regions every time a vortex occurs. A tight vortex caused by our smaller sun line is a tornado, and they only seem to occur in the northern hemisphere.

The smaller vortexes of tornadoes reoccur in relatively the same region and only occur during the day in that region of our Earth which is facing our sun. It is important to understand that they do not occur at night because our sun lines do not point onto our Earth at night. It is also important to remember that the ley lines which the tornadoes are reacting with are constantly on the move due to the north on the compass always being on the move.

A cyclone which is about 1,000 times bigger than a tornado occurs every twelve months when the neighbouring line is pointing onto our Earth. They are generated by a sun line which is 1,000 times bigger than our sun. These cyclones have been reoccurring and following the same relative paths in the same relative areas for as long as we can trace our history, but most importantly, if you study the path taken by a cyclone vortex, you will notice that the cyclone has a tendency to meander during the day under the momentum of the powerful vortex when the neighbouring line is not pointing onto that region of our Earth. The same can be said for hurricanes and typhoons which also follow the same relative path of the coastline, but I have mainly studied cyclones because they're in the local news here in Australia, where I live.

The cyclone is a vortex of much greater destructive diameter and strength than a tornado, and the cyclone continues to spin under momentum during the day. During the night when that same neighbouring sun line is pointing onto our Earth in that region again, the cyclone appears to regain activity and speed at night and all too often follows a designated path very close to the coastline. Remember that our Earth is on a tilted axis. When it is night-time, the region of our Earth which is facing the neighbouring sun line is now being subjected to a cyclone every twelve months in that region that is in the same polarity, thus the opposing reactivity.

During the day, the same neighbouring sun line will be on the other side of our Earth and still pointing onto our Earth, but it is pointing onto the opposite polarity which is not reactive. It meanders during the day (under momentum) because the cyclone is not overwhelmed by the neighbouring sun line during the day. When the cyclone occurs, it gathers fine dust particles which are in the opposite polarity and are also charged up with an attracting energy in the opposite polarity to the line. Please note that it is only the particles which are in the opposite polarity. Because of opposite polarity particles, the cyclone moves towards the place where the sun line is being projected onto the Earth again during the night when the particles are attracted to the sun line.

Cyclones in Australia can sometimes occur in the north-east coastline or in the far north-eastern region of the Coral Sea. My theory is that this is where the neighbouring lines are injecting into that region of our Earth for that type of energy and vortex.

Remember that there are also two different sizes of neighbouring lines. I have noticed that cyclones can occur in either the north-east region of Australia or further out in the eastern region of the Coral

Sea, about 1,000 kilometres off the east coast of Australia but also in Africa. All locations are in an east–west or longitudinal direction to each other. The neighbouring sun lines are also in two different sizes and lengths. The longer neighbouring sun lines will deflect further and will not follow the coastline like a shorter neighbouring line will do. When the longer neighbouring line is projected onto our Earth and a cyclone is generated, it will occur in the far eastern region of the Coral Sea after the line undergoes the deflection of 1,000 kilometres out to the Coral Sea.

My theory is that each solar system takes four years to complete one full rotation. This means that the magnetic sun lines projected from the neighbouring solar system will be pointing in this direction every twelve months at a quarter of its rotation period. Cyclones, typhoons, and hurricanes are recorded every twelve months and in relatively the same area, and this is when the neighbouring sun lines will be projected into our solar system onto our planet.

Let me also make an important prediction here. If the sun lines from our sun and from neighbouring suns have a polarity to them as well as a different length, then this means that each vortex will be different but will reoccur under a set of sequences. By a set of sequences, I mean that the atmospheric vortex will be identical and will repeat itself every four years when that same or identical sun line enters our Earth's atmosphere again.

Because the sun lines exist in different lengths, this will cause each sun line to project onto a different place on our Earth due to the amount of deflection. This variation in length will cause a different deflection of each line as they rotate around each sun, especially in the length and deflection of neighbouring sun lines. My prediction is that cyclones, hurricanes, and typhoons which are generated by the neighbouring sun lines will reoccur in the identical path or pattern, including intensity **every four years**. This is because the identical sun line is generating the same cyclone, hurricane, or typhoon every four years. It could be that a cyclone in Australia is generated one year, followed by a typhoon the next year (around the area of Japan) and a hurricane the year after (in the Americas) and then another cyclone the year after that but in the region of Africa.

Let me describe a little about the mechanics of an atmospheric vortex with the finite details later in another chapter, if I haven't already done so. We already know that an atmospheric vortex is caused by eddy currents, but eddy currents are caused by the sun line being in the same polarity of the Earth in that region. The atmospheric vortex (large or small) is caused by a repelling reactive polarity caused by the same polarity.

We also know that our Earth has north and south polarities. The neighbouring sun lines and our sun lines have a north–south, south–north polarity. If a cyclone which occurs in the southern hemisphere is affected by a south polarity sun line, then it only goes to say that, the next year, a cyclone will occur again in the southern hemisphere because the next line is a south polarity. The next line is, of course, a north polarity and will cause either a hurricane or a typhoon. The following year, the same hemisphere will also be affected. BUT the length of neighbouring lines which cause cyclones, hurricanes, and typhoons are at different lengths. This means that the large atmospheric vortex will occur in a different ocean each time it is generated.

I would like to make a safe bet and say that a typhoon near Japan will occur one year and a hurricane the next year or maybe the reverse in the same region. ALSO, remember that the neighbouring sun has different-sized lines, and this means that different deflections will also occur but ALSO different-sized atmospheric vortexes. This beckons the question 'Is a typhoon more powerful than a hurricane because both of these vortexes occur in the same hemisphere or vice versa?'

It also beckons the same question about Cyclones. Is every cyclone of the same strength in the same area? It could be that a cyclone occurs during one year in Africa and another the next year in Australia, but the cyclone in Australia may be more intense. Remember that the neighbouring sun lines are also in the same configuration of north–south, south–north.

A typhoon may occur around the area of Japan after the second cyclone during the next year, when a neighbouring line is projected there in a different polarity. A hurricane may occur the next year in America when the same polarity of a neighbouring line rolls across our Earth as the sun line which generated the typhoon in Japan. The neighbouring sun lines are also north–south, south–north.

The neighbouring sun lines also exist in a large north and then a small north, large south and then a small south. The large north line may generate a typhoon around the coast of Japan (in the northern hemisphere) but may have occurred in this area (in the east) due to the large deflection of that line. The small north line may generate a hurricane (in the same hemisphere) in the Americas and may have occurred in this area due to a small deflection of that line. The next line, which is a large south line, may generate a cyclone (in the southern hemisphere) around the coastline of Australia or out in the Coral Sea (east of Australia) due to the large deflection of the south line. The next line is a small south line, with the next vortex off the coastline of Africa due to a small deflection.

Let me also say that (in its early years) the Earth had come off its paternal line, and the crust was very soft. During these early years, our Earth—as Gondwana—broke up into many continents caused by the gyro effect, plus the further shaping by these lines. After our Earth settled into its own orbiting pattern, it was then subjected to the sun lines of our sun and of neighbouring suns. During this period, the sun lines were very strong; and when the strong neighbouring sun lines were projected onto our Earth, they cut and reshaped the continents into the final shape they are today.

Today after many millions of years, the neighbouring sun lines are still projected here onto our Earth, but they are now weaker in energy and are only the cause of cyclones, hurricanes, and typhoons. (Today they are no longer strong enough carve up the continents.) The fact that these neighbouring sun lines shaped our continents is the reason that these cyclones, typhoons, and hurricanes now follow the coastlines today, and the cyclones that may occur in Australia are an excellent example of this. As previously stated, I have noticed that a great deal of hurricanes throughout Hawaiian history, dating as far back as 1843, have occurred on or about the fifteenth and sixteenth of August (https://en.m.wikipedia.org/wiki/List of Hawaii hurricanes).

These sun lines have a numerical projection, and I have also made accurate predictions of earthquakes that are caused by the sun lines in the opposite polarity of our Earth, using a calendar when counting the days, hours, and minutes from one quake to the next. The way that I did this (which was also with a witness) on one particular day during 2011 was whilst working for the local city council library. I asked the librarian to look up **when** the last two major quakes had occurred in New Zealand. I then measured the exact amount of time from the first quake to the second (in days, hours, and minutes) and forecast the third quake and also the fourth using the same amount of time. There was also a very insignificant fifth quake. My calculation was accurate within one hour of the third and fourth quakes over a number of days, and all these quakes occurred in a Z pattern for the four significant quakes. Plus, the information in previous chapters stated just three strong lines, but there were five quakes because there were actually five significant lines which were strong enough to affect our Earth at the time. One of those lines, however, was only strong enough to generate the fifth as an insignificant quake.

Later, I describe how the projection of sun lines are in patterns on our Earth in much the same way that a tennis ball has patterns on it. The sun lines are projected all over our Earth. This makes me wonder about the tectonic plates. It may be that the tectonic plates are also lines on our Earth which were impressed on by the neighbouring sun lines that are only 100 times larger than ours. These tectonic plates are also in synchronised tennis ball patterns, but they did not cut into our Earth as strongly as the neighbouring sun which is 1,000 times bigger. If we study the eddy currents caused by the sun lines, we may possibly learn a lot more than what we have about eddy currents from within the electrical industry.

Tornadoes occur more frequently than cyclones because tornadoes are generated by our sun lines which project onto our Earth every three months because our Earth is orbiting our sun. However, the neighbouring sun lines which generate cyclones, typhoons and hurricanes are only projected onto our Earth every twelve months because one rotation of sun-star systems takes four years. A tornado is caused when a sun line from our sun is projected onto our Earth which is in the same polarity as our Earth for that region. Remember that our sun projects its sun lines onto our Earth every three months.

The energy which is generated when the sun line meets our Earth's ley lines is a repelling energy, and the incoming sun line twists around in a reaction with the ley lines, generating an atmospheric vortex. The reason the atmospheric vortex is generated is caused by the flexibility of the incoming sun line which causes it to spin because the ley line is fixed into the Earth's crust. Because magnetic north is always on the move (along with the ley lines), there will never be a tornado twice in the absolutely same location during our life cycle.

The particles spinning inside the vortex/tornado are spun around due to the attracted polarity of the particles. (The particles in the opposite polarity within the atmosphere are attracted to the sun line.) The tornadoes are only generated when a repelling energy is present, and they do not occur in that location or area when the very next king tide occurs because the next sun line is in the opposite polarity. Remember that the tornado spins due to the repelling energy of the ley line.

During the very next king tide line, the energy is an opposite polarity or attracting energy, and this generates a volcano eruption or associated earthquake. The volcano eruption does not occur in the same region of our Earth as the vortex due to the deflection of the next line. (More on volcanoes soon in a later chapter.) These sun lines are also the cause of sinkholes, and I'll describe the mechanism which causes those soon as well.

Earthquakes

Earthquakes are resonant vibrations from bouncing the mantle up and down. I have described in good detail (I hope) the theory of how our Earth began as merely a light frequency that grew stronger during billions of years.

I also mentioned earlier that each core of our Earth is a result of each inner tube, generating a separate frequency which initially began from a light energy. I now say that the cores exist in separate frequencies because each tube has its own frequency of energy, and each core will therefore exist in resonance with its paternal tubed line. My proof of each core existing in an individual frequency can be seen when we look at exactly how a quake is generated.

Remembering exactly how timed all our galaxy really is, we can clearly see that every solar system is taking four years to make one rotation and that two timed neighbouring sun lines crossed over to create our sun-star. The timing of neighbouring sun lines and our sun's lines still exists today. In the timing of sun lines, a line from our sun and a line from our neighbouring sun meet head to head, with our Earth in between, during synchronised events. When the lines meet head to head, an intensified energy is generated. The neighbouring sun lines created our sun, and this mechanism occurred in a domino effect. Every sun is therefore in resonance with each other. The sun lines from our sun created the cores of our Earth, and those cores are also therefore in resonance with every sun line.

Under the basic laws of magnetic attraction and repulsion, when these lines meet (which are in resonance with the cores), they cause the mantle to bounce or vibrate due to the resonant vibration which causes an earthquake. (Remember the violin string and the piece of paper; it really is that simple.) When a quake occurs, it is only that particular core of our Earth which was created by that particular tube line (as just one of the tubes of a full line) which is generating the resonance vibration whilst the lines meet head to head. Just as the vibration that occurred to the string of the violin caused the piece of paper to vibrate, so too does the core when it bounces.

Lightning Strikes

This reveal on lightning begins with a small mention about a recent number of extremely powerful lightning bolts that have seemingly come from out of nowhere as they are reported in the media. These are usually referred to as 'dry strikes'. One particular dry strike killed many livestock, and another incident occurred where a single bolt of lightning struck a yacht, which ignited. I saw both of these reports just this week (2018).

I will describe how the sun lines cause lightning strikes, and the above is clear evidence that the injecting lines are intensifying during this period after the planet alignment. I'll describe the cause of that intensity again.

The majority of lightning is generated when a sun line is being projected onto our Earth. Let me say that I have noticed that the majority of lightning during the day is more of a bolt of lightning, whilst lightning at night is usually sheet lightning (spread across the sky). This is due to the intensity of our sun lines projecting onto our Earth during the day when our Earth is facing our sun. The magnetic lines from our sun are stronger in energy than neighbouring sun lines, and sheet lightning is generated by the neighbouring sun lines which aren't as intense. The sun line passes through a cloud that is holding a static charge, and the sun line then injects into our Earth and grounds the static charge.

As we know, the air between the cloud and the Earth ionises under the electrical laws of ionisation, and the atmosphere is therefore placed under pressure when it becomes the medium that helps get the line to the ground. Before it does, our Earth's atmosphere, which is in the light blue band of the frequency of blue, does ionise before it gets to the ground. When the lightning bolt travels through the air, it is following the path of the magnetic sun line which is leading as the bolt travels towards the Earth.

When a polarised flash of electricity occurs in the laboratory from a high-voltage device (maybe from one brass ball to another) to the other object which has been earthed, the flash of electricity is doing the same thing as the lightning bolt in nature. (It is following the path of magnetic line energy.) It's a well-known fact that lightning does not always strike the highest point, and this is due to the injected sun line not being injected into that precise location (where the tallest object is positioned) but rather into the exact location where the powerful sun line and lightning strikes.

https://blog.gotopac.com/2013/04/17/air-ionization-how-it-works. This link shows you the cause of ionisation and you might like to click or copy and paste this link into your search bar or search for another link that describes ionisation.

Lightning, as we know, is millions or maybe trillions of volts which have built up inside the cloud due to static. The magnetic line flowing through the cloud gives the static charge a path to Earth, and it is logically at the same frequency of the magnetic sun line along with the crust of our Earth.

Recently, I saw a video on social media of a lightning bolt at night which rose out of the Earth and forked across the sky. This type of lightning occurred due to a magnetic line from our sun pushing its way through our Earth and into the sky in the opposite direction of the average lightning bolt. It would not surprise me if this lightning bolt in reverse was one of the king tide lines from the diagram and that it occurred in the hour of the king tide.

But the list of lines that can project or inject onto our Earth as lightning is incredibly long. There are lots of open-circuit and closed-circuit lines from our sun and from two different neighbouring suns.

This brings me to my next subject before I teach you about my theory on the mountain ranges.

It always rains after a line passes by our Earth whilst orbiting. The best manner in which to convey this reveal is actually told as a story. I was at the Gold Coast twenty years ago, which is a little south from where I am now in Beenleigh, Queensland, Australia. I felt the urge to go over to the water canal close to where I was parked in the car. I stood at the edge of the man-made water canal on top of the storm water drain that fed into this water canal.

I looked out into the canal and saw a large sandbank of gravel coming out of the storm water drain and curving outwards to the ocean in the same direction as the outgoing tide. I concluded, due to the sandbank curving out towards the ocean, that it very obviously rains a high percentage of the time during an outgoing tide. This logically means that it rains, most of all, after our moon and its magnetic lines have passed or after one of the many sun lines has passed over our Earth.

With this in mind, I have watched our Earth tentatively throughout the last twenty years for just about any storm or rainy season I could witness. The rain has been more intense every three months during the change of the seasons or during king and neap tides (especially of late) because of the magnetic lines being induced with neighbouring line energy during their synchronicity.

On the third of May 2018 in Hawaii (around the time of a king tide), a huge volcano erupted.

The volcano eruption had to have been generated due to an attracting reactive energy which draws out the lava whilst the passing of the three strongest lines was occurring. (I'll have to describe the term 'draws out' properly in the later chapter on volcanoes.) Around three months later, on 26–28 August 2018, a massive hurricane and flood occurred in Hawaii. This would have been due to the sun line in the diagram of that period passing whilst being in a repelling energy. The volcano around the time of a king tide was an attracting energy, and the hurricane three months later would have been a repelling energy.

The hurricane was not precisely three months later, but remember that there are actually three strongest lines that we orbit through every three months, when one of the diagram lines pass by our Earth. Therefore, the volcano in May could have begun at the start of the first strongest line, the eruptive momentum then continued over the next coming months, and the hurricane could have then occurred anytime during the period of the next three lines, causing the eddy currents to probably occur during the last line for that period. The hurricane (named Lane) occurred on 26 to 28 August and could have been caused by the last of the three strongest lines for that period. There were actually four hurricanes in Hawaii during that particular season, but the most intense was Hurricane Lane with a slightly weaker hurricane before and after.

I have told you about the coat hanger (Cox meter) which was rotating at an immense rate during August 2018. During the last month now (as I write the final pages of the book) in October 2019, I have watched a number of videos in social media reporting extreme weather conditions. A couple of those were viewed just last night that also recorded an immense amount of hurricanes and extreme weather, including flash floods in numerous countries, plus extremely intense lightning bolts. I've described throughout the book that our Earth orbits through one of the four lines every three months. I have only just now (2020), months before publishing this book, watched videos and found a list that begins in 1843 of hurricanes in Hawaii. The hurricanes in Hawaii occurred mostly on the fifteenth and sixteenth of August, and I can tell you that it is not simply hurricane season. (https://en.m.wikipedia.org/wiki/List_of_Hawaii_hurricanes).

Another very good example that I found had occurred in Italy during 2018. The list below is a very good example of an attracting energy occurring in Italy during one king tide period and then an opposite repelling energy occurring during the next king tide period.

- volcanic eruption - 24 August 2018
- floods - 7 October 2018
- floods - 26 October 2018
- floods - 31 October 2108
- floods - 4 November 2018
- floods - 1 August 2018
- floods - 4 July 2018
- floods - 31 June 2018
- floods - 9 May 2018
- floods - 14 March 2018
- floods - 13 December 2017
- floods - September 2017
- floods - August 2017
- floods - January 2017

- floods – 26 November 2016
- floods – 20 November 2016

Now if you look at the volcano eruption on 24 August, it would have been the most resonant energy, thus the most intense. If you look at all the dates before the volcanic eruption, you will notice a gradual increase from the year 2016, with the floods gradually increasing through to November 2018.

I must point out that floods occurring in 2016 had longer dry periods than the next flood, and this is a good indication that both the lines are injecting into that area of the globe and also that the line is getting closer to resonance, thus the strongest intensity. There has also been intense flooding in Russia and in China during 2018, and they're all indications of sun lines being projected onto our Earth, especially when you look at the dates of the flooding.

In a chapter soon, I will be describing how a sinkhole energy is generated by a repelling line energy, and there have been incredibly intense sinkholes occurring in Russia, which I will describe in detail in that chapter. There was a bridge collapse in Russia (2018) around the same time as the floods there, and this was coincidentally because the huge concrete bridge structure was right in the firing line of a sinkhole injecting line energy.

All this extra activity on our Earth is not the coming of the world's end; it is simply intensified sun lines, whose cause I will further explain in later chapters. Briefly, all this is due to the remnant energy of our planet alignment, plus the synchronised event of a neighbouring planet alignment. Science recently announced the discovery of that massive planet at the outer edge of our solar system, and this is an indicator of immense energy lines being projected into our solar system from that neighbouring one.

The flooding that took place in Hawaii after Hurricane Lane is due to intensified lines passing by Hawaii just as the sandbank had shown. There has been both effects experienced in Hawaii that I have taken heed of. I will describe to you soon how the magnetic energy throughout our entire solar system is ever increasing (apart from the strong points made throughout previous chapters).

What we must now know is that there is actually a more technical reason that it is happening now which has to do with the planet alignment. This increase in intensity of weather was caused by an energy recharge during a frequency alignment (planet alignment). In short, the extreme weather conditions have become more extreme after the alignment. I asked you previously, Do you think weather has intensified after the alignment? We can, of course, naturally expect it again on the next planet alignment in AD 2500, but I don't think it will be as intense due to my theory of a stronger energy generated every 2,000 years by a neighbouring alignment, which occurs at the same time every 2,000 years.

I have noticed strong indications that every 2,000 years, neighbouring solar systems are also aligning, and the energy is greater during their alignment. Those neighbouring lines are doubling up with the energy of our lines during periods of synchronicity and amplification. Considering the doubling up of the neighbouring sun lines with our solar system lines and the equation of alignment energy is very real and is an extremely strong energy.

Sinkholes

The basis of a sinkhole energy is that planets generate a Hall effect when a neighbouring line is projected between them. This link shows nearly every major sinkhole which has occurred around the world: https://www.pinterest.com/karlarlloyd58/sinkholes-around-the-world-incredible/. The link below is a video of a recent sinkhole that occurred inside a Russian city which was built on top of a natural

potash deposit: https://www.rferl.org/a/russia-sinkholes/28890114.
html. Remember that I previously described that one line generated and formed our Earth in a frequency that was in resonance with carbon, and my greatest example of carbon for you today comes in the form of potash.

During our Earth's creation, the magnetic energy must have become extremely resonant and somewhat stronger in the frequency of carbon in the region of our Earth where that city was built. Every sun line is in the same frequency; the evidence of that is that every sun attracts the same atoms, and therefore when neighbouring lines inject into our Earth (as they have always done so), they are in the same frequency as our Earth after it was formed by a line of our sun. Our neighbouring suns entered and crossed over, creating our sun, and I realised that synchronised projections by neighbouring lines will continue to occur today as they always have, especially by the identical resonant line in that same polarity that created our Earth in that area.

In my studies at college about electronics, we were taught that a very small magnetic signal can very easily be amplified by a mechanism identified as the Hall effect. The Hall effect in electronics generates a stronger component of electrical energy from two magnets which are side by side. The way that it's done is to place two magnets side by side with a fine gap between them and to then pass a disk possessing a magnetic spot on it through the centre of the two side-by-side magnets. The magnetic spot on the disk constitutes movement within a magnetic field, which is the primary law of generating electricity; and during the passing of the magnetic spot on the disk, the Hall effect generates a spike of electricity which is induced into the two side-by-side magnets. The energy spike generated as the spot on the disk passes through the two side-by-side magnets is sent as an electrical signal through a cable into an electronics-sized transformer, and the output then becomes an amplified wave in the identical frequency (under the laws governing transformers). The most common use of the Hall effect in electronics is in older cars from the '90s.

The following describes where the Hall effect occurs in our solar system. The first stage of the Hall Effect occurs in our solar system when orbiting before and after a planet alignment. As previously stated in earlier chapters, the planets will begin to form loose groupings and then tighter groupings before the alignment and again in mirrored positions after the final event. During these stages, the planets will move from loose groupings to tighter groupings. A tighter grouping of the planets will simulate the generation of a stronger inducement energy to the magnetic spot passing through the two side-by-side magnets. The magnetic spot is simulated by the magnetic sun line from a neighbouring sun which is passing through two side-by-side planets.

The planetary Hall effect that I'm about to describe in detail is weaker when the planets group loosely and very intense when the planets are (more or less) side by side. I know that the planets are never actually side by side, but when the neighbouring sun line travels through the gap between planets at light speed, they (more or less) appear side by side to the sun line.

I have identified the planetary Hall effect as the 'Cox effect' for ease of identification throughout the book. Under the Cox effect, a neighbouring line or our sun's line passes through the gap between two planets, and an extremely amplified energy is generated when the line passes through the gap and exits out of the other side of the two planets which are relatively side by side. This occurs in almost the same manner as when the pool wave is amplified into a fine spout of water which shoots up into the air under my description of resonance.

The actual mechanism of the Cox effect is like this. I mentioned previously that every magnet has the same frequency as an unbroken or un-refractured magnetic line. Every un-refractured magnetic line is a combination of every light frequency on the EM spectrum, and therefore, every magnetic line exists in the identical accumulation of frequencies. The two planets creating the Cox effect are pulsing out an identical physical wave which is in resonance with the approaching line which is about to pass through the gap between the planets. At the gate of the gap, the incoming line enters, driven by the force of momentum.

When the line approaches the gap, there is a physical wave of magnetic energy from each planet which is pulsing into the gap. This pulse is a force against the incoming line at first, but the momentum carries the line through. When the passing line enters the centre of the gap, the side-by-side waves compress the magnetic tubes into a tight circumference (like the fine water spout), and the momentum of the line and the synchronised pushing of the side-by-side waves shoot the passing compressed line out of the Cox effect, and it continues towards our Earth in a compressed state.

When the two Cox effect planets are in loose formation, the passing line is not as compressed, and the result is also a loose set of magnetic tubes or a wider circumference than a passing line that has passed through two planets close together as in the above description. You may have noticed that some sinkholes are not as intense or as deep and appear to be over a wider area of the land. (This is the greatest example that I can think of for a wider and weaker sinkhole.)

I theorise that the weaker sink-holes are caused by the weaker neighbouring lines. I must also add that these wider weaker sinkholes seem to be more apparent on coastal regions due to the synchronised timing of orbit, rotation, and movement of neighbouring lines. Remember that these neighbouring lines once carved up the outline of the continents, and this is why the region close to the coastline is hit by sinkholes, especially here in Australia. The mechanism of the pulsing is that every wave is identical, and the two side-by-side waves combine their force to turn the line into a compressed line and shoot it out of the Cox effect like a slingshot.

I also mentioned earlier that the planet alignment is also a recharge and frequency alignment. During the planet alignment, all the frequencies of the planets become synchronised. After the alignment, they pulse at precisely the same frequency. During an alignment, when the planets pass one another, the magnetic energy of a stronger planet induces and recharges the energy of each nearby weaker planet. Before the alignment, the planets form loose and then tighter groupings. After the alignment, the energy is stronger and the planets form the same groupings but in mirrored positions.

The recharge and frequency alignment increases the intensity of the Cox effect. After a planet alignment, the Cox effect is more powerful, and you will notice that the sinkholes are more intense. For you to completely understand every generated Cox effect line which injects into our Earth, I need to briefly show you the following facts about electricity supply and then come back to the sinkhole mechanism.

Paralleling Of Two Identical Waves Of Electricity Supply

In the electrical industry, before two sources of electricity can be connected in parallel, the wave of each source must first be synchronised. The waves must be identical.

The way that we did it in the distribution network when paralleling or adding a generator into the grid system was to use a special electrical meter that displayed both waves and then synchronise the two waves before we connected them. The same mechanism is used when adding the input of solar panels (on top of the houses) into the grid system.

Sinkholes Continuation

Now what's incredibly important to know is that, before a planet alignment, after 500 years of orbiting, the frequency of every planet is slightly out of sync, and the frequency alignment adjusts that factor. Therefore, before a frequency alignment, the waves of the two Cox effect planets are slightly out of sync, and the resultant compression of the passing neighbouring line is not as tight or as amplified in formation before a frequency alignment. In fact, the resultant compression of the passing line before an alignment is quite loose or weaker and wider. After an alignment, the two side-by-side waves involved in the Cox effect are in absolute synchronicity, resulting in a very intensely compressed line which is very powerful.

The next stage of a Cox effect injection of the compressed line is for the line that is in resonance with the cores of our Earth to want to join with the resonating core for being in the same repelling polarity with our Earth in that area. The compressed sinkhole line, for being in resonance, begins drilling into our Earth's crust as a result of the resonant vibration whilst trying to get to the mantle which is of the strongest resonant magnetic energy. Further, the outer core and the materials of the crust are also in resonance with the injecting magnetic tube line, and the materials help with the vibrating and drilling.

The magnitude of sinkholes varies in their circumference and intensity of the incoming energy due to three main factors (that I may have already mentioned), which are the loose grouping, the actual magnitude of the waves of each Cox effect planet, and the recharge frequency alignment. There is an incoming sun line which is amplified by two planets which are side by side. The line is compressed into a fine spout of energy which is then injected into the crust of our Earth in a repelling energy and (more or less) drills its way into the crust. The result of the injection of some sinkholes in some areas of our planet has turned into a stroke of bad luck for some particular locations, where the energy was injected and the sinkhole occurred.

I think there are two particular injections which I've noticed that were exactly in the middle of concrete bridges. One occurred in Russia, where the huge bridge collapsed in the middle, and another I saw just last night in a video of another concrete bridge in Italy in 2018. I have mentioned both of those areas previously. One area has had a volcanic eruption, and both have had massive flooding, which is also a strong indicator of the presence of passing sun lines. Also, just this morning, 8 November 2019, the news reports massive building destruction in areas of France, which is most probably caused by loose injections. I must also add that the report also mentioned poor stability in those buildings due to poor construction techniques throughout that area.

The injection of compressed sinkhole lines has been occurring for as long as history can record and is somewhat random due to our Earth continually orbiting and spinning 24/7 and the sun lines having to be synchronised as well. During the past billion years, the neighbouring lines have grown in size and in their level of energy. The reactive energy within the centre mass of each sun has grown immensely in mass and reactivity, which reflects the lines of those suns today.

Let me add that the two planets being side by side does not occur that often outside the period of a planetary alignment, and the same is true for sinkholes. But tornadoes occur all too frequently. In closing, sinkhole injections have occurred during alignment periods more often than when our Earth is orbiting in between planet alignments. They do still occur anytime but more frequently during alignment.

Mechanism of the Pool Wave Spout and the Light through a Magnifying Glass

The following may also help understand the strength of the pulsing which is generated by the two side-by-side planets. To help visualise the mechanical pulsing of two side-by-side planets, consider the stone being dropped in a pond when the water waves pulse out from the centre. If you can imagine two of these stones being dropped into the water side by side, it might help visualise the two planets side-by-side.

It may also help to remember the windows of your car being open whilst in transit. Whilst driving with the rear windows open, it will at some point generate a pulsing on your ears when the two wind waves are generated inside the car (normally felt coming from the back of the car). These wind waves are sometimes quite powerful, and just now as I drove down to the local shops, it was suddenly generated at its strongest inside the car at exactly 55 kph. The wind wave was, of course, always present but reached resonance at 55 kph due to the collision of the two wind waves coming in each side of the car. The windows were open at just the right size of the opening and the car travelling at just the right speed.

What I am about to describe is a huge fact to comprehend. The poles of our Earth emit the pulse of 8–12 Hz, but the crust emits the frequency of blue light energy.

Remember that every magnetic line exists in the form of multiple concentric tubes and that the neighbouring lines are enormous today. Science records every planet with its own frequency, but I believe what they are measuring is the frequency of the light energy. Remember what I told you about our Earth's pulse and its other frequency of blue light energy. Both are different, and the same is true for every other planetary frequency recorded. Remember that every magnetic energy is an electromagnetic line of frequencies which combine together to pulse at an accumulated frequency. The energy of every magnetic line that is unbroken or un-refractured is therefore in resonance with each other—absolutely every one of them.

Gouging and Carving by the Amplified Lines

Every gorge and canyon in the world was once created by an injecting line which travelled across the landscape and gouged out the crust which was in resonance with the line. Two different types of gorges were carved into the crust during two different periods. The extremely narrow sandstone gorges that lead to the city of Petra are a prime example of how narrow the beam of the amplified injecting line became when generated and then thrust into our Earth millions or billions of years ago.

Some time back, I saw a satellite photo of the sandstone rockscape surrounding that nearby area of Petra, and it clearly revealed many lines carved into that area of sandstone which all led back to one central point like the apex of a V. The Grand Canyon is an older and wider carving of our Earth because its gorges are wider and in harder rock. I theorise that they were created when our Earth's crust was at its softest, after it was broken up into the separate continents. I also surmise that the gorges of Petra were gouged out from our crust after all the planets were orbiting, after they had carried out their first alignment, when the planets were in frequency synchronisation, which would have helped.

I actually refer to the carving up of gorges and canyons as part of those tennis ball patterns on our Earth. The gouging was logically after Gondwana broke up into separate continents, but that wasn't the last of the carving up by these lines. If you study the globe with the gouging effect in mind, you will notice that a lot of the continents form common or familiar shapes or carved designs. The American and Indian continents are very similar in their carved shape. Australia has been trimmed up at the sides and is somewhat like a diamond in shape. Its closest neighbour is a strip of land that was carved off the side of Australia and became New Zealand. Tasmania must have also been severed off as a lump that was once attached at the bottom of Australia.

Let me tell you that the gouging of the crust and the shaping of the continents were carried out by a repelling energy line that was trying to get to the inner core, just like a sinkhole line. When an injected sun line is in an attracting energy, it occurs during a different orbiting period. Because of our orbiting and spinning every twenty-four hours, the continents received a number of repelling energy line/s that generated large indentations into the crust.

One good example of an indentation into the crust is in the very centre of Australia as Lake Eyre. It's normally dry and only fills during an occasional flood. Lake Eyre was indented into the crust of central Australia when the injecting line was in a reasonably dull energy with a very wide circumference of the same polarity as the area of our Earth in Australia. The Lake Eyre indent was caused by the injecting line (more or less) rolling around that area. The indent was not the size of the line. This indent occurred during a period when the crust was softer.

There is also an estuary which was also gouged out into the Australian landscape that travels from South Australia all the way up and into that area of Lake Eyre. It is about 1,800 km long. **Remember that it rains heavily into flooding when a sun line passes an area.** This estuary, named Cooper Creek, is said to be about 115 km (70 miles) wide during extreme flooding, which has only ever occurred twice in the last 220 years obviously when that same line that created the lake and estuary is passing by Australia. Why they refer to it as merely a creek, which is this wide during flooding, I'll never know.

Volcano's

Mountains are extinct volcanoes that were sucked out of our Earth during a doubled-up attracting energy. The basis of this theory is that volcanoes are generated by an amplified attracting energy which causes the lava to be sucked out of our Earth which is the result of one neighbouring projected line doubling up with one of our sun lines. Let me try to show you how.

Before we start, I need to refresh your memory of the violin whilst being tuned in. You will need to remember the slightly weaker vibration when it was almost in tune and the perfectly resonant frequency, which was the strongest vibration, when it was perfectly in tune, and then there was a repeated almost in-tune vibration which was generated on the aft side of the perfectly in-tune vibration. I described how the lines generate sinkholes, but that is a different amplification which repels. The only real difference is that sinkholes are a resonant repelling energy that must drill through the Earth's surface.

Volcano line energy is a doubling-up, amplified attracting energy that sucks out the lava. Let me try to show you how it occurs. I need to also bring your attention to the fact that 99.9% of all extinct volcanoes now exist as mountain ranges, which can be described in ley terms as 'lines of mountains'. The mechanics of extinct volcano ranges is very simple; after I describe the ranges in lay terms, we'll then look at the technical theories.

Extinct Volcanoes As Mountain Ranges

Extinct volcanoes always start from one end of a mountain range as small mountains that were created from a slightly weaker resonant energy line which were generated from an almost in-tune attracting or sucking energy. The mountains increase in height until the middle of the range, where the highest mountain exists, and were generated by a perfectly in-tune resonant injecting line energy which was (at the time) the strongest attracting or sucking energy. Then after the highest mountain, which always appears to be in the middle of the mountain range, the height of the mountains gradually reduces again, which were generated by another almost in-tune attracting energy which was slightly weaker again, where the range height then dwindles away even more. The greatest example of this highest middle mountain is in the mountain range that surrounds Mount Everest.

Now let me show you how the attracting or sucking energy occurs. When a sinkhole energy is generated, it drills into our Earth and is caused by a line which is amplified under the side-by-side planetary Cox effect when injected into our Earth in a repelling energy. The side-by-side planetary Cox effect doesn't occur as often as a volcano attracting energy, and that is why sinkholes aren't in the numbers equal to that of mountains; plus, the volcano energy which causes volcanoes (mountains) is much more powerful than a sinkhole energy.

The mountains as volcanoes were created when the Earth's crust was very soft. The volcanoes which eventually formed as mountains were created and erupted after Gondwana broke up into many continents. Look at the largest sinkhole which is very old that occurred in Russia; it was caused by a repelling energy line. Compare it with the sucked-out result of Mount Everest with its great number of surrounding mountains through to hills.

Volcanoes were created by a reactive amplified energy under a **doubling amplification effect,** and again, the lines that generated volcanoes were in an opposite polarity energy to a sinkhole line when volcano lines were in attracting polarities. The doubling-up amplification is generated during extremely synchronised events when the neighbouring lines projecting into our Earth occur at the very same time as our sun lines are projecting through our Earth.

Remember that the neighbouring lines were once long enough to create our sun and today are a great deal longer; plus, they created our sun lines and are therefore in resonance with them. All sun lines exist in the same frequency. They are in resonance. Our sun lines which are in tubed lines generated our Earth, especially its cores, so cores of our Earth are also in resonance with tubed lines of neighbouring suns. The neighbouring sun lines and our sun lines rotated in synchronicity when our sun was first created, and the neighbouring lines and our lines are therefore extremely synchronised. During their rotation in the period of early Earth (when it was extremely soft), all sun lines were straighter, and this gave them the potential to be very intense under this doubling-up amplification.

The mechanics of the doubling-up effect is generated like this. The neighbouring sun line projects onto our Earth at exactly the same time as our sun line is projecting through our Earth. Each sun line is coming from each side of our Earth, and during this time, they meet with our Earth in the middle in an attracting polarity. Our sun line was always stronger, and the result is that our sun line, with the neighbouring sun line (when doubling up), helps the neighbouring line attract and suck out the lava.

What also occurs during this doubling up is that our sun line (more or less) helps in the sucking out by pushing the lava at the same time. (That description really needs a full chapter to explain, but just try to theorise it for yourself at the moment until I can make a video.) Because of every line being in resonance, especially the neighbouring line being in resonance with each core of our Earth, the doubling-up attracting energy became extremely powerful. The only way I can assimilate this energy is to observe the double bounce on a trampoline.

Our sun line, coming from that side of our Earth, projects through our Earth at the precise moment that the neighbouring sun line also projects onto our Earth, and they meet head-on with our Earth in between each line. During their rotation whilst meeting, they both generate an incredibly strong doubling-up attracting energy and still do today but not as often as when all the volcanoes existed on our Earth. Remember that when the volcanoes were first created (extinct volcanoes are now mountains), the lines were straighter and stronger; plus, our early Earth was much softer. The energy was a great deal stronger; the Earth was a great deal softer, making it a great deal more possible.

Recently, there were countries that also suffered flooding due to the lines passing across our Earth in those regions, and it was also because every in-tune line was very strong and more in tune after the planet alignment. Today our sun line is a great deal more powerful than the neighbouring lines, and our Earth is in resonance with both the neighbouring lines and our sun lines.

Again, our sun line energy (during a volcano energy) helps push out the lava whilst the neighbouring sun line will be attracting the lava out. This is what I call doubling up, and the result is a volcano eruption. In the early years of the then softer Earth, the lines sucked out lava from our Earth after repeatedly meeting and doubling up for millions of or a billion years.

The result during this period of 'stronger lines' was that active volcanoes turned into inactive volcanoes, becoming mountains after that period. The result of the lines doubling up today is very rare volcanic activity, but the results are there in plain sight. The result nowadays is more along the path of an earthquake rather than an active volcanic eruption.

More Mechanics On Volcano Lines

When both of these sun lines (the neighbouring sun line and our sun line) begin to meet at the start of this doubling-up cycle, they do not begin by meeting in the centre of each line. The slightly weaker lines (out of the three strongest lines) which are on the fore side of the strongest line begin to meet first, and it is a slightly weaker attracting energy. The result today is that the affected area suffers the beginning of volcanic activity which just rumbles, generated by resonance. The lines are only almost in tune when this occurs, which is not the strongest attracting energy (or sucking energy); plus, the core bounces due to resonance, causing quakes in the area.

Then as our sun and the neighbouring sun rotates (along with our planets), they then repeat the projecting lines and this cycle of volcanic energy. The doubling-up lines occur again as they have done over millions of or a billion years. The neighbouring sun line synchronises its doubling up again with our sun line as the perfect frequency gets closer; therefore, the attracting energy gets stronger. When these lines meet and double up, it occurs in a similar way to the tennis ball patterns which gouged out the gorges and canyons, except that the doubling up is in synchronised injections that are an attracting or sucking energy by the lines.

The greatest example of doubling-up line occurring is when, whilst Gondwana grew as stagnant Earth, it was growing within the tube line frame. Therefore, whilst our stagnant Earth rotated around the sun (on just one of our sun's solar system lines), the doubling up was totally impossible. (I hope that you can visualise that.) I am saying that there was a total absence of sucked-out volcanoes during the stagnant Gondwana Earth period. Volcanoes were ever present during this period, but that was because our Earth's crust was thin, and the lava wasn't far from the surface.

Volcanoes and the doubling up of lines only began occurring after our Earth was pool-balled. After being pool balled-off the line and due to our Earth's crust still being extremely soft during the early Earth Gondwana period, doubling up was ever more possible. This is good evidence, and I will also add that sinkhole injections under the Cox effect were also totally impossible in Gondwana due to the absence of other planets orbiting during this stagnant Earth period, plus the fact that our Earth was growing within the frame of the magnetic line.

Actually, right now (while I think of it), mountain ranges are in a similar pattern (in lines) to gorges and canyons, so maybe the doubling-up volcano lines occurred at the same time as the repelling gouging lines. I guess we could step back and take a look at extinct volcanoes (mountains) all over to see if they're in a straight line or slightly S shaped, like the gorges, but I just haven't actually had that thought until now, just hours away from my deadline of publishing.

My sincerest apologies if you're sitting there at the moment thinking, 'Well, I'm totally lost.' I promise that I won't leave you like that, There will be a video which will make it much easier to understand.

I am about to explain why extinct volcanoes begin as hills, grow to small mountains, progress towards being the highest mountains, and then dwindle away again in size during that 'line of mountains'. During our early orbit after being pool-balled off the line, the volcano attracting lines were getting closer and closer to the strongest, most perfectly in-tune, or most resonant attracting energy which was also vibrating the crust and inner core (also causing quakes during volcano eruptions), which generated the most reactive volcano and the highest peak.

The synchronised injection of a weak line and then a slightly stronger line, a stronger line, an almost strongest line, another stronger line, the strongest line, a slightly weaker line, a weaker line, and so on is the actual cause of the extinct volcanoes being in a line as ranges, with the middle being the highest. It was two lines doubling up which generated the result of an extremely powerful energy in comparison to a side-by-side Cox effect, causing a sinkhole. Plus, volcano lines occurred more often because the lines rotate and meet more often during doubling up than a Cox effect.

There are, of course, some single extinct volcanoes, now mountains, seemingly in the absolute middle of nowhere at times (like Ayers Rock in the middle of Australia). I've given it some thought of late but can't figure out why these somewhat random volcanoes occurred other than to say that synchronicity is somewhat mysterious. Ayers Rock is in the centre of Australia and was once a mountain said to be 10,000 ft high and is just one of those single extinct volcanoes. It is only 700 kilometres from Lake Eyre and Cooper Creek, which fills Lake Eyre during each rare flood.

Cooper Creek is about 1,800 kilometres long, flowing from South Australia all the way up to the centre of Australia to feed Lake Eyre. This is said to have been 115 kilometres wide during the last two occasions in a 220-year period, and this was due to the three paternal lines passing by, causing massive downpours. My theory says that these lines also created Cooper Creek and Lake Eyre.

You can hopefully see that volcanoes were a result of lines which doubled up, and there were none of these types of volcanoes on stagnant Earth, as Gondwanaland, because our Earth was rotating or orbiting whilst affixed on the paternal line. Today each of the four lines is curved and has begun to be extremely reactive of late because the planet alignment is complete, the resonance is absolutely perfect, and therefore, the energy of late is the strongest. Our Earth and also the lines (therefore our entire solar system) are increasing in reactivity, thus flux density, but the lines are still curved, and this makes them less powerful.

You will also hopefully see by the date of the Hawaii volcanic eruption in 2019, which occurred over a number of days (due to injecting lines meeting in powerful resonance), and the date of the hurricane three months later indicates that the lines are very active every three months. Plus, the hurricane was not exactly three months later because the perfect frequency of resonant lines generated the volcano eruption but was slightly out of sync when the hurricane and floods occurred. I hope that you can also see that the severe flooding occurred under my theory of the mechanism where it rains after the lines have passed (remember the sandbank coming out of the storm water drain) and that the hurricane/flood in Hawaii was severe because the volcano showed it was a strong line energy. This occurred a great deal more when our Earth was softer as early Earth and our sun had straighter lines, and this is also why most volcanoes have been extinct for millions of years.

Let me show you some more very good evidence. When there is an earthquake, very often, there is an ooze of mud that rises out of the Earth's crust and science refers to this ooze as 'liquefaction'. A very good example of this liquid earth or liquefaction occurred in Christchurch in 2011.

I am actually only theorising again because I haven't been to Christchurch to inspect the ooze, but I am going to go out on a limb here to say that the liquefaction or ooze of mud which came up in Christchurch actually was probably very magnetic. I often wonder whether the magnetic liquid earth would have also been helping amplify the total of magnetic flux. So again, could someone help us all out here and see if they can find some remnants of the liquefaction in Christchurch and check it for magnetic energy?

I can definitely tell you that the injecting line which generated the volcano in Hawaii was of the opposite polarity and was therefore an attracting (sucking) energy, and if the hurricane occurred three months after, then this hurricane line was most likely of the same polarity and an opposing energy line.

Solar Flares

This is a theory which only came to me the day before finishing the manuscript of *The Holy Grail of Science*. It was prompted by a friend who mentioned solar flares. That person told me that solar flares were occurring quite frequently of late, back in November 2019. My friend also had bought a book on electromagnetic pulses (EMPs) exactly three months prior when the last batch of solar flares had occurred.

Solar flares, as we know, are eruptions of reactive materials that pour out from our sun's reactive mass. Our sun was generated by a crossover of lines which were projected from two neighbouring suns, and if the lines were huge back then, they're obviously way beyond humongous today. Every sun's lines are incredibly synchronised, and I can tell you that neighbouring lines still project into our sun today.

When a solar flare is occurring, it is a neighbouring line doubling up with one our sun's lines; and just like the volcano attracting energy (which sucks out the lava), the lines affect our sun's reactive matter. Solar flares are reactive matter being sucked out of our sun's reactive mass by the doubling-up energy of overpowering neighbouring lines every three months.

The electromagnetic pulses generated by 'sun flares' are not increasing our sun's radiation; it is the neighbouring sun's lines rolling through our Earth during the solar flares. The EMPs that they proclaim are causing electronic/satellite dropouts are caused by the doubling up of lines on our Earth. No amount of protective equipment will stop EMPs from occurring; they're way too powerful.

My theory of the lines crossing over jumps into medical science. Science recently announced the discovery of a 'flash of light' that occurs in that nanosecond during the moment of conception.

Science reports say that, from that moment of conception, the cells seemingly divide.

I can tell you that the cells are not dividing, but they are projecting rotating energy lines out, just like neighbouring sun lines did when they created our sun and crossed over. The magnetic energy lines of these cells are creating little magnetic pumps when the cell seemingly divides. Cells, as we know, consist in a number of atoms which form to make molecules, and those molecules combine to form the cells. When the cell lines cross over, they attract atoms, which form those molecules that eventually form the cells and so on under the same laws of magnetic resonance; they quickly create the next new cell, which in turn creates the next cell and so on. I can tell you that atoms contain moisture, and the cells are immersed in this fluid which accumulates to moisten each cell. Science recently announced that each planet is highly likely to possess water, and this is because of rule number one: atoms contain moisture.

Here's my theory on ocean swells and storms at sea which create the huge ocean waves. This theory may be a little harder to conceive because there has been such a strong description by science associated with ocean swells for such a long time before I offer you my theory now. I believe we have always been misled about the source of the ocean swells, and that will therefore make it difficult to accept my theory here. It is also very new and it begins like this.

All magnetic energy exists in the identical frequency, and when all the planets align, it is also actually a frequency alignment. This theory on oceanic swells began for me, while I was thinking about the ocean swells and the fact that our Earth very obviously has a frequency of blue energy which also pumps out a pulse in the band of blue. The pulse of blue energy is emitted from the crust, where it projects out from there in a physical wave and a visual display. It does actually affect our oceans by way of its ocean swells. I was thinking that maybe the pulse wave of the core does actually generate our oceanic waves when I realised the following.

The ocean in many regions is very deep. The deeper the ocean is, the closer the water is to the source of the 8–12 Hz pulse; thus, the pulsing of the core generates the ocean swells by pulsing the water. The frequency of the pulse, however, is a great deal higher than the frequency of ocean swells. The only way that I can describe the reason that the ocean swells are at a lower frequency is to tell you a short story which helps explain this.

I was driving down the freeway one day recently when a car next to me presented an optical illusion where the mag wheels of the car (at just the right revolutions per minute) appeared to be motionless. I concluded that the pulse of our Earth in the frequency of 8–12 Hz, which is much higher than the frequency of the oceanic swells, has (more or less) the same effect on our ocean waves as the mag wheel has on my eyes. I also remember thinking in the early hours of that morning that some lakes are enormous, but they are nowhere near as deep as our oceans, which means that they are not affected by the pulse of our Earth in generating big waves on them. I contemplated that sometimes (especially during storms) the ocean swells become waves which are huge, and I then remembered that sometimes there are other planetary energies from those other planets in our solar system, and their energy is being induced into our Earth's pulse. Together with the other planetary energies, the pulse of our Earth is doubling up with the pulse energy during storms and adjusting the frequency of the ocean swells to become huge waves.

Induced energy into the crust of our Earth causes increased pulsing of the oceanic swells when they become waves, whilst these other planets (in our solar system) are passing by our Earth; plus, maybe neighbouring sun lines are generating huge storms. Our sun's lines, our other planets, and our moon, plus the neighbouring suns, could all be amplifying and creating the differing frequency of our

Earth, thus pumping the ocean swells. So larger ocean waves may be generated by stronger magnetic frequencies, and differing frequencies may differ the ocean's frequency of the waves. I'm just saying that, surely, the wind doesn't generate every ocean wave, but also, what about huge lakes, especially those in the United States? If the wind generates waves, then why not huge waves or swells on those lakes.

The following is actually another example of a resonant frequency expressed in the sound barrier. It may be quite a few years before science will actually recognise any of my theories in *The Holy Grail of Science*. I mentioned in a previous chapter that every frequency of sound exists in the form of waves and irrefutably has a direct relationship to every frequency of electromagnetic waves.

When a fighter jet breaks 'the sound barrier', a cloud of moisture appears which is in the frequency of the white energy that is generated, and this is the combination of every visible light frequency. It is a moist cloud because it is created by a vacuum, and moisture is affected by a vacuum. Every frequency of sound travels in a wave. The barrier of sound is the combined frequencies of every sound just like white light is the combination of every visible colour frequency. At a particular speed (around 767 mph), the jet breaks that physical accumulative wave; thus, a physical vacuum is suddenly generated. Therefore, every frequency which includes light, electricity, magnetics, sound, vacuum, and velocity is all directly related to one another (https://www.businessinsider.com/fighter-jets-breaking-sound-barrier-2016-3).

Bird Brains

I'd like to explain what I believe happened to all those birds that died in the thousands during 2011 and 2013 in America, Canada, Australia, and other parts of the world. Some of you will remember this as it was so sad but baffling at the same time. The facts are that, during 2011 and 2013, there was an incredibly huge number of small birds—but mainly parrots and small blackbirds—that were found dead on the ground and had simply just dropped dead in Australia, America, Canada, etc. This was due to a neighbouring magnetic line being projected into those areas of our Earth in a frequency which was in perfect resonance with the brains of these small birds, causing intense fatal migraines. To help understand how this could have possibly physically occurred, it will help to go back to the chapter on resonance and look again at the 'butterfly vibration'.

In considering magnetic lines, I also ask that you consider a dog whistle which is only heard by a dog and realise that just because we can't hear a sound doesn't mean that it doesn't exist. I also believe that migraines suffered by some people are also caused by planetary frequencies combined with our moon.

Contaminated Waters (Creeks, Rivers, Oceans)

This is not a statement about the effects of magnetic lines, but it is related to the following explanation on whales and dolphins and must be brought to the attention of others. We need to stop and take stock of the state of our creeks, rivers, and oceans. I saw an oily sludge floating in the tributaries in local parklands which were open storm water drains fed from nearby factories. The creeks local to my city feed into the river system, which is nowadays void of any real fish where the majority of breading is supposed to occur. The ocean where this local river feeds into is incredibly filthy around the mouth, and fishing there has dwindled off to just about nothing.

It's no longer possible to take a drink from contaminated creeks or rivers, and it might sound ridiculous to even think of mentioning this, but drinking from these tributaries was once a common event within my father's lifetime. I visited a friend about a decade ago in a brand-new housing estate near me, and in their front yard was a water tap marked Do Not Drink. It was treated sewage water 'on tap' in these front yards that was not a fenced-off area, out in the open. Those taps should be locked, but most importantly, stop and look back for just one second and think about how 'accessible drinking water' has evolved too.

So there may be lots of suburbs like that now with a ready supply of treated sewage water. Then there is the appearance of that luminous or bioluminescent green substance that you see when you kick the sand at night, and it glows in the ocean waters.

Whales and dolphins are beaching, and fish are dying by the ton. Dolphins are beaching due to depression, which I believe is caused by a build-up of mercury in the oceans, causing mercury poisoning, as well as all the other pollutants. Some years ago, I had all the mercury amalgam fillings removed from my teeth; and for a period afterwards, I suffered severe paranoia for no apparent reason. I put it down to all the mercury which spilled over into my body and flooded my system during that removal.

Just think for one moment what all those ocean mammals are going through. I read just last night (25 November 2018) that there was a reported 140 tons of dead sardines that appeared on the beaches in Chile (the actual amount is a bit ambiguous), and it was in the north. The reports were that it was due to a massive algae outbreak that choked up the water. Exactly three months before the sardine incident, there was a severe earthquake in the south of Chile, and I can tell you that the quake was the cause of a magnetic sun line, and three months later was when the next major sun line was due again.

The massive algae outbreak was caused by the passing of a major sun line. The energy of a magnetic line temporarily increases intelligence (as I have described below in the chapter of 'evolution'), and it's my theory that the magnetic lines projected into those areas (where the algae occurred) and caused a sudden evolution of the algae. The description about the sardines choking up in the water occurred because the major sun line that our Earth orbited through created the algae.

In Chile (the southern tip of that continent), there was over 300 whales that beached at around the same time that a magnetic sun line was due to project onto our Earth. It is my theory that the whales suffered extreme migraines (the same as the birds but at a different frequency) and had to beach to escape the pain. (As depressing as it may seem, the death of every whale isn't caused by being full of plastic bags). DON'T GET ME WRONG, THE BAGS MUST GO **NOW.** REFUSE THEM.

In New Zealand on 10 February 2017 at around the event of the king tides, whales and dolphins had beached again whilst this February line was being orbited through, but I must say that quakes often occur in New Zealand.

The Social Effects Of Magnetic Sun Lines

The following items are (more or less) my own studies of the effects caused by the lines on society. The planetary alignment generates a very strong energy which is, of course, stronger than the energy of general consensus that is generated for centuries before and after an alignment. From what I have concluded, the stronger energy begins to truly increase around 120 years before an alignment occurs and stays until about 120 years thereafter.

In the last 120 years, there has been two particular greats within the science world, who were, of course, **Tesla** and Einstein, but there has been a number of people of notoriety who lived during the energy surge of a planet alignment. During the last alignment in AD 1500, the most memorable were Da Vinci, Nostradamus, Galileo, Henry XIII, and others. Jesus Christ, as we all know, was reported 2,000 years ago, and the Gautama Buddha was born around AD 567, but Confucius was also born in 551 BC, which means that they all lived across a planet alignment.

The amount of effects by these lines are colossal, but I'll tell you a few other direct effects that everyone is totally unaware of. The most important fact that I have to tell you is that the effects are by way of energy on our bodies, plus our minds; and for some, it affects us socially or mentally.

If you haven't already figured it out, the lines cause social stress, and your outlet of stress is dependent on where you are in society. By this I'm saying that, from the first day that I saw the diagram of four lines, I almost immediately began to notice robberies, murders, and increased violence in the community both in my local community and in the media around the world during king tide and neap tide events, and I've had to painfully stand by watching it until I could inform you of it in this book.

A great number of animals navigate their way around our Earth by using the energy of ley lines. During the last twelve years, I've noticed whales and dolphins beaching, and science has reported it as depression. Science is totally naive to the intense energy of the lines upon the minds of these animals, plus the disorientation (caused by the incoming sun lines) when confusion sets in on their natural minds' navigation system which blinds them from their usual ley line navigation.

I also began noticing that orienteering people were getting lost in the bush in various countries during king tide days, and I felt that it may be due to their compasses pointing in slightly different directions whilst they take their reading from the compass (just like the occurrence of rotation of the clothes hanger). I've recently contemplated that these people who become lost whilst orienteering may be like some animals using a 'compass mind'. Nearly every species of animal uses the 'compass mind' and ley lines.

There is definitely a science to people who bushwalk. They are very compass-oriented people in their minds. Some may have been using a directional compass mind and gotten stressed out whilst the magnetic sun lines were passing by, but I'm sure that you'll agree that they were very likely dehydrated.

Recently, I saw a very good example of dehydration in the report of a man in the local news. He was lost on a little island just outside our major city. It was only a small island, but he became dehydrated after becoming lost. He was stressed out, getting more and more lost, and then total disorientation set in. He then suffered from extreme dehydration, and within about four to five hours, he died.

Consider a man who walks in the desert and becomes dehydrated. There are a number of stages of dehydration. I personally know because I've been through life-threatening stress and have been dehydrated badly. The stages are of stress and then panic, resulting in further dehydration (I learnt from the extreme example of that man on the island), and then disorientation and uncontrollable confusion. Your mind can play horrible tricks on you, especially thinking that it is the end for you, maybe seeing just a mirage, or having just extremely negative thoughts.

I have also noticed during sun line injections on king tide days that some people who have little experience in flying a light aircraft have crashed during nights of high energy and high levels of rotation by the Cox meter (clothes hanger). This may be because they were flying after just attaining their flying license, and then during high rotation of the hanger, they lost the accuracy of their compass gauges on the dashboard or something and crashed through lack of experience due to gauge loss. I call this 'the John Denver and Shirley Strachan theory', which is how both of these men were killed. They were both new at flying, ran out of petrol after becoming lost at night, and died in a crash.

Please let me say that if you are a person who becomes stressed at times (maybe due to these sun lines), helpful and healing modalities are a good option, and I am thankful to those in that industry. Primarily, music has been such an incredibly healing gift for me and to those who also make it, plus the multimedia industry, including cinema and TV. They have also given me great joy and healing at times of need, and I thank all those who produce it.

Finally, in regard to social effects caused by the overall frequencies of planets (as mentioned briefly on the butterfly vibration), sometimes the positions of planet groupings can generate emotions. Every emotion (remember, the word 'emotion' describes energy in motion) exists in an individual frequency, and the greatest example of a frequency generating an emotion is the monk who strikes a gong and makes every effort to attune his body and mind to its frequency, which exists in the frequency of peace.

In the long line of frequencies of sound which are available, there is a vibration which is in resonance with the emotion of peace. I must point out that love is just next to peace on this line of frequency emotions. Let me show you a timeline in society which will show you that love and peace are next to each other. If you look back to the period forty years before the alignment (around the 1960s), you will remember that this period was renowned for an extremely strong or overall expression of the love emotion generated by the group positions of some planets before the alignment. So sometime after the early '60s, the overall emotion of the general public shifted from the emotion of love slowly towards an eventual overall emotion of peace.

In closing, I would like to say that some people are more affected than others when it comes to certain types of emotions, which are generated by planetary frequencies, and I can't quite describe the cause of that in this book. A video of this needs to be made. The list of social effects really is quite extensive, but I will leave it at that.

THE EVOLUTION OF
THE SPECIES

Before I present this theory to you, I'd like to point out that everything is affected by magnetic energy. You can look into social media sites to see a simple block of wood that is floating on water which is repelled when a strong magnet is placed near it. I concluded through my studies that the foundation or the primary cause of evolution is the 'tipping of the scales' when an increase in the component of electrical energy comes in force. How does that occur on our Earth and in our solar system? Well, the answer to that really is quite simple.

During the period of Gondwana or stagnant Earth, our Earth did not move very fast through the solar system, whilst it was affixed to the paternal line. I am reminding you now that our Earth, at one stage, revolved around our sun, affixed to that paternal line, but only made one complete revolution every four years. I will also remind you that the scales of electromagnetic energy will favour the side of electrical energy when there is movement within the magnetic field. As the coil of wire begins to 'cut flux' lines, this will, in turn, generate electron flow, and the scales will tip. This is the primary rule in generating electricity or electrical energy.

Another rule on increasing the output or electrical energy is by increasing the flux density or the strength of the magnetic field within which the coil is rotating. As our Earth rotated on the paternal line (affixed to the line), in the period of stagnant Earth, it did not move through a great deal of magnetic energy and did not cut a great deal of flux lines. But the magnetic field did lightly strengthen over that period as our sun became more reactive, increasing the strength of the energy field within our solar system, thus increasing the amount of electrical energy. Remember that our solar system is completely embroiled in magnetic line energy with many closed-circuit magnetic lines, and the all-important king and neap tide lines radiate from our sun, but during the period of stagnant Earth, our Earth did not cross any magnetic energy lines whilst being affixed to one of those king tide lines as the paternal line.

During the period of Gondwana, there were originally just four species of dinosaurs in one massive ocean which had evolved to eventually also walk on land. The four original species merely began simple life as a generated source of energy in much the same way as our sun was created as a magnetic pump, except the original four were generated in one massive ocean with the help of electrolytes from the seawater.

As we now know, electrolytes (especially Celtic sea salt) will help overcome the side effects of chronic fatigue. The cells, as previously mentioned, become discharged, causing chronic fatigue, and the sea salt will recharge those cells. The neighbouring lines cross over at a point within the waters of that ocean, creating mini-magnetic-pumps in the ocean, helped along by the electrolytes, and this was the manner in which the four species were generated and eventually evolved. Remember that I previously described the way that the cells seemingly divide. When they eventually evolved out of that Panthalassa, they finally became land-dwelling dinosaurs. It all began during Gondwana, when the ocean was rich in electrolytes. Also, the Celtic sea salt was moist; in the packet, it said that it contains microbiotic organisms which we consumed before we evolved out of the ocean.

I can't explain exactly how it all first occurred, but those first four species which evolved in the ocean were there due to four frequencies being projected into that ocean by neighbouring sun lines. The Earth was rotating upon the paternal line at a very slow rate compared with the period after this when Gondwana broke up into many continents. Rotation around the sun every four years fixed on the line and never spinning or rotating as a globe meant that our Earth was subjected to only a very small level of movement within the magnetic field of our solar system and consequentially low levels of electrical energy. These first dinosaurs which existed on Gondwana were reported with a very low intelligence, and I say it was due to very little movement or the generation of very little electrical energy as the 'evolutionary energy'. There was, however, eventually just enough movement and increased magnetic flux to produce enough electrical energy for the first four dinosaurs to evolve out of that ocean and to eventually walk on land but with no great numbers or other species to evolve, unlike the next form of dinosaurs.

You may now truly see why I refer to Gondwana as stagnant Earth. When stagnant Earth as Gondwana was pool-balled through space, our Earth began to rotate as a globe every 24 hours and eventually slowed down to begin its own orbit every 365 days. With rotation of the globe every twenty-four hours, the evolution of the species also increased, especially in the species of plant life which already existed. Those first four species of dinosaurs were almost wiped out during that first pool-balled ice age, but some survived; my guess it was from being in the parts of the Panthalassa which didn't freeze over or being deep in caves.

We must also remember that the crust was also very thin and soft during Gondwana, with a large number of volcanoes simply bubbling up out of the thin, soft crust. The temperature of our Earth around this time when being pool-balled through sub-zero temperatures was already very hot from the volcanic activity caused by the thin crust, making it very volatile in comparison to the later 'early Earth' with many continents. In an earlier chapter, I described Gondwana as a stagnant Earth with a massive continent which (more or less) faced into the wind whilst it grew on the paternal line with a massive ocean called Panthalassa. For those dinosaurs to have survived (I am only guessing), when our Earth was pool-balled through space, it left the warm atmosphere behind, and the one land mass of Gondwana was still facing into the wind when it was pushed through space with the ocean behind it and out of the wind.

A good way to confirm this is to throw a tennis ball through the air which has a heavier mass fixed onto one side of the ball. The tennis ball will obviously turn into the wind or oncoming path, with the heavier mass leading the way, as the tennis ball floats through the air. Remember that the area of space surrounding our Earth today (which is actually heated by our sun) exists at sub-zero temperatures of around -300 degrees Celsius.

After Gondwana was struck and pool-balled through space, it then began to break up into the continents which still exist today after it rotated continually every twenty-four hours and was subjected to a gyro force on a soft, new-born Earth. Our Earth then slowed down from being pool-balled to a rate of orbit of 365 days for each yearly cycle and also began rotating as a globe every 24 hours, which also meant that there was a great deal more movement within the magnetic field of our solar system. Our Earth was now also rotating within the field of every closed-circuit line which existed within the entire solar system and was also crossing through the king and neap tide lines every three months. This was an entirely new ball game for our globe. (Pardon the pun.)

Evolution is the cause or side effect of a greater component of electrical energy. When Gondwana broke up, a greater evolutionary force was generated by a greater movement of the globe within the new magnetic field after beginning its own orbit every 365 days and continual rotation every 24 hours. Evidence of my theory was seen in the next species of dinosaurs which roamed the Earth on the many continents where and when a much higher intelligence was recorded by science, including the remaining dinosaurs from Gondwana which would had also evolved again. But where had this new species come from?

Recently, a NASA space shuttle returned to Earth, and science discovered bacteria on the windows of the craft. NASA then concluded that space is apparently full of bacteria. They also discovered recently that Mars has a very poor protective atmosphere and is also inhabited by the same types of bacteria. After Gondwana was struck by Mars, it was then pool-balled through space and eventually slowed down and thawed out from the effects of the first pool-balled ice age, and the flesh of the rotting dinosaurs attracted the bacteria which may have either come from outer space or from Mars when it hit our Earth. My theory here says that Mars struck the Earth and then pool-balled it through space. It may have also been this contact with Mars which gave the bacteria the opportunity to inhabit our Earth, or the bacteria may have gathered on Earth whilst travelling through space (whilst being pool-balled). It may have also been due to both of these reasons that our Earth was inhabited with bacteria.

I'm sure that from this theory on evolution, you can visualise some of those bacteria eating the rotting flesh of those dead dinosaurs that would have eventually evolved into meat-eating dinosaurs but also the bacteria which ate rotting vegetation which would have also eventually evolved into dinosaurs as herbivores, plus every other different type of dinosaur in between. With our Earth now orbiting of its own accord, it was then subjected to a greater number of frequencies during its orbit, which would have started the process of evolution all over again; but whilst being submitted to a stronger field of magnetic energy, the potential now existed for a greater number of bacteria to develop, which is exactly what occurred. A stronger magnetic field projected onto our Earth, which now existed with many continents and then produced a greater number of species with higher intelligence.

Every one of the first dinosaurs evolved from energies when stagnant Earth was subjected to neighbouring sun-star lines crossing over in the sea, and the next species or forms of dinosaurs evolved from bacteria except for those leftover dinosaurs which made it through the first pool-balling ice age. The many species of today have evolved from the dinosaurs, which came from space as bacteria and have evolved further since the last ice age. We also need to look at the list of peculiar creatures discovered in our oceans which were boosted by the electrolytes (that may have helped them evolve) to realise the strength in that statement.

There were a greater number of species that lived on the many continents after Gondwana, after the majority of the first species were snapped frozen during that pool-balling ice age. The next species in their varied numbers were now orbiting at a rate which was four times faster than Gondwana within a greater magnetic field, also rotating twenty-four hours a day. This meant a higher percentage of electrical energy was present, and the new species of dinosaurs were said to be of higher intelligence after being subjected to a greater force of this evolutionary energy.

I previously mentioned (in an earlier chapter) that the evolution of the spider occurred around the same time as when Gondwana broke up into many continents. Can you imagine for a moment that, after stagnant Earth was pool-balled through space, it was then subjected to a new magnetic field which was much stronger as Earth moved at a higher speed through space whilst being pool-balled; thus, a higher percentage of electrical energy was subjected to those scorpion like creatures on our Earth which became spiders. Those spiders evolved at a significant high rate during the break-up of Gondwana, when there was a greater movement of our Earth through space and the magnetic field of the entire solar system. The evolution of the spider occurred during the period when the Earth was being pool-balled through space, which meant there was excessive movement until the Earth slowed down.

Since the period when many new species of dinosaurs had appeared, there had been other ice ages which almost wiped out these new species. Since those other ice ages, the magnetic field has increased in strength and number of lines over millions of years. There was then the introduction of many species which were more intelligent as a result of the increase in the field strength of the magnetic energy within our solar system. The increased amount of field lines would have been a primary ingredient of increased intelligence.

You may have heard some people talking about a 'higher vibration'. We are continually moving into a higher vibration where I believe there is a greater force of evolutionary energy. This higher force of evolutionary energy is also coupled with brain activation caused by sugar and other stimulants. A higher vibration again has been felt of late whilst the entire world overindulges in stimulants, very much like the hummingbird who lives on sweet nectar.

I'm sure that you have understood my theory that there was minimal subjection in the early days by the electrical component with my supporting evidence, that is, Gondwana had scorpion like creatures that evolved very quickly to become spiders. But the most important point is to tell you that our Earth, during this period, was subjected to an extremely low amount of magnetic energy as well, and there was very minimal movement which could generate a cutting of flux, which is the primary principle of creating electricity. Throughout time, the flux density within our solar system has been gradually increasing as the magnetic component, thus increasing the electrical component during movement in orbit.

Just as it did during the last alignment (during the Da Vinci period), there have lately been an incredibly huge amount of new technologies and inventions. The alignment energy begins 120 years before the event and stays with us for 120 years after the event. This can be especially gauged from those scientists who lived 120 years ago. I have not really studied who lived 500 years ago. The period of 1960 up to today was also very strong in evolutionary energy. Let me show you the evidence of that.

Science records the migration of cane toads here in Australia whilst they spread to our beautiful Kakadu National Park, and the evolution of that specific species of the cane toad has accelerated extremely fast after the toads figured out that they needed speed in their stride to cover the wide-open spaces. The toads, now hopping into Kakadu, have been recorded with longer legs after evolving to this new species in such a short amount of time. This was due to our Earth being subjected to an extremely strong magnetic density during the planet alignment, and that includes the influx of energy from our neighbouring solar system. That huge planet which has been discovered at the edge of our solar system during 2018 is a strong indication (as previously mentioned) that the neighbouring system is also aligning in our direction, which has increased the level of energy in our system.

You might be asking, 'How come specifically spiders and cane toads?' Well, the easiest way to explain that is to say that the energy being projected onto our Earth was also in their resonant frequency.

The Evolution Of Specific Birds And Other Animals

Someone had pointed out to me a TV episode of a lyrebird being photographed in the wild, which was over a period of a few months (I guess). During that relatively short amount of time of being photographed, the bird reportedly began to imitate the camera shutter noise. I also saw a video about twelve months ago of a magpie bird that was imitating a dog barking and other noises. (You might have seen it.) There was also another recent video of the same species of bird that was actually a better example of evolution. This shows that there are a number of birds in that species which are evolving at an accelerated rate on their way to becoming a new species.

Also, recently, the animal welfare people in zoos have announced that a greater number of varied species (**NOT the same species**) are also responding to their images in mirrors. Today, 26 October 18, in the news here in Australia, there was a koala which was enjoying his connection with his trainer so much, which is a good example of this species evolving. I can't forget the most incredible example in the animal kingdom, where Koko the Gorilla used sign language. (Seeing Koko the Gorilla is a must for all ages.)

There are also an incredible amount of internet videos of incredibly intelligent dogs, horses, and cats. Dogs must have been the very first animals to personally interact with humans, closely followed by horses. All those are very good evidence of evolution both in nature and socially with humans.

Physical Human Evolution

We are all rapidly evolving both mentally and physically due to the internet and medical science, especially with genetics but also with human robotics or orthopaedics. In the last 100 years, all science has evolved at an incredible rate. I believe, in some areas of evolution, it may have been that medical science was a result of the gymnasium additives. Some gymnasiums offer a lot more additives outside the normal parameters than they should, but I personally feel they shouldn't be inciting or encouraging anyone (especially the youth) to indulge in them. Vitamins and other stimulants, like coffee, are also pushing us to evolve both mentally and physically, but they're not all bad or used in a bad way.

Children are recently recorded as being born with six fingers, but this is not a new concept. They had recorded six fingers and toes on the giants that roamed our Earth thousands of years ago. You all should look into all the internet videos and reports about absolutely every living creature on our Earth, not just the possibility of giants—EVERYTHING. You will only believe what you are ready to believe, and the same must be said for *The Holy Grail of Science*. It is all just a theory.

THE FOLLOWING IS A STATEMENT THAT I WAS GIVEN FOR ANYONE READING THE BOOK.

If **Your** beliefs tell you to take another's life. Then you need to immediately change your beliefs.

If you are committing genocide in the name of your flag then stop immediately and also re-think.

Put down your nuclear weapons or you too will only suffer from the side-effects of a detonation.

If you are thinking about children in a sexual manner then you are ill and you need help immediately.

If your belief is that you can leave destruction on our planet, then you also need to change.

If you are an Authority and are breaching the Law yourself, you must stop.

If you are manipulating the Court or Laws because of your belief, then you are doing wrong.

If your belief is that **your** medicine is needed more than another's then you also need to change what you believe. (https://www.disclose.

tv/all-found-murdered-doctors-who-discovered-cancer-enzymes-in-vaccines-333465)

In reference to the last message, we all need to give greater credit for the abilities of natural healing being practised throughout the world (i.e. all those who do spiritual, faith, energy, or any other type of natural healing).

Here's just one example that was given to me by an Australian Aboriginal lady that you can try. We nearly all suffer from leg cramps during sleep, and the cure for a high percentage is to simply follow this remedy. Place a cake of soap near your calf muscles under your bottom sheet.

The following is a short story of my life leading up to *The Holy Grail of Science*. These were the events which led to my first miracle when I called to God in an arrow prayer, but I must warn you that my life didn't start in the proudest of circumstances.

Just in case you didn't know, I was told that an arrow prayer is the most powerful prayer request that you can make to God. It was said, 'An arrow prayer shoots straight to the heavens and into the very seat where God resides.' The simplest arrow prayer is said when you look to the heavens and call HELP!

The story of how I arrived at that moment went a little like this. I was a hard-working man from the first day of starting work. I'm sure that there are others who worked much harder, but I've always put in every effort possible. My father and mother led very good examples of this, and both of my brothers today are workaholics, who have also been through immense change.

Anyhow, I bought my first house when I was just 20 here in Australia; and a month after that, I turned 21 and bought my first Harley-Davidson, a 1952 Hydra-Glide. It honestly and sincerely wasn't the bike purchase or the bike itself which brought trouble in my life. It was the attitude that was around in that era of Harley-Davidsons, which is very different today.

As a hard worker who loved to ride, I also bought another Harley that I rode to work and on weekends, a Harley Softail Custom. This was the silver-blue model with a running writing of '*Harley-Davidson*' on the tanks. I loved it, or I should say 'it loved me'. I rode hard on weekends and drank way, way too much beer, getting into pub brawls and becoming a bit of a Casanova. (Well, that was what I thought.) I never ever went near drugs, although late in this period I was pressured into some rubbishy stunt using marijuana and a bucket, but it was just one puff of indulgence (probably curiosity more than anything else).

When I look back on my life in the few years after turning 20 until about 27, it turns out I was just a beer-mongering, pie-eating pig, and that is honestly how I view myself today during that period. However, if you were broken down on the side of the road or were having problems that I thought I could help with, I did so.

The following is an example of myself back then when a Spanish family lost their 13-year-old daughter after she ran away from home. Back in those days, I would simply listen to my gut. I had that young lady home that next day within hours after simply going to the city and asking around in the street community.

But my somewhat Christian approach to life was not the view of a girlfriend who decided to 'take me down'. For whatever reason I would never conceive, that was exactly what she did after she started up on hard drugs (cocaine and heroin). The story that unfolded as a result of this cathartic woman's mind went a little like this, but just for ease of reading, let's refer to her as 'Cathartic Cathy'.

When I first met Cathy, I was about 25 and had just begun working very hard as an electrician on wages for a firm that was a contractor to the construction and maintenance of all the Sizzler restaurants and Kentucky Fried Chicken outlets owned by Collins Foods. I wasn't working for this subcontracting company for very long when, I guess, I had passed all the tests after jumping through the hoops. They then asked me if I wanted to work under subcontract and travel to the remote cities doing the same work, and I agreed.

I hadn't really been going out with that cathartic woman for very long when I met her parents, and still to this day, I don't know how I missed the signs. Her father and mother turned up at my house, where Cathy and I had started living together. It was solely my home (well, the bank owned half of it), and that was (more or less) how Cathy introduced me to her parents. 'Hey, Mum and Dad, this is Mick. He owns a large part of this house.' I'm honestly not kidding; that was how she introduced me.

Looking back to those days, it is very obvious that Cathy had never had a boyfriend who was so responsible and was also worthwhile of introducing at such an early stage of the relationship. Was she a gold-digger? I'll let you decide, but some of you who have the experience may have already heard enough.

Cathy worked for Social Security as an investigator on cases of fraud, busting the butts of those who were cheating the system—well, at least trying to, until they met Cathy and her crew. I came home quite late after completing overtime the same afternoon or, should I say, the same night as I received that offer of subcontract. I had to tell Cathy that I was leaving for Cairns in the far north of Queensland in the next couple of days. Cathy said, 'Right, I'm coming with you.' She went into work the next day and gave the government an ultimatum that they let her go with me and give her an immediate transfer, or she was taking six months off to be with me. Cathy received her immediate transfer and was in my van with me, headed to Cairns just two days later.

I had driven straight through from Brisbane to Cairns in a matter of about twenty hours, which was usually a two-day trip of 2,000 kilometres. The boss was extremely demanding and told me to rent a house with extra bedrooms so that he and his wife could come for a supervising holiday at some stage of the project, all paid for, out of my own pocket, of course. The rent in Cairns was extremely high, being the second biggest tourist spot in Australia. I rented a house a little out of town from an announcer of the local radio station.

I arrived in the same afternoon at the job site to see what my biggest challenge was and then headed home for a good night's sleep. I woke very early the next morning full of nervous tension. Cathy was giggling in her job; she was being paid living-away-from-home allowance and was, of course, also living out of my pocket. I had no idea exactly what the etiquette was on that situation.

Within the first week of arrival, I'd hired some local labour to help me out. I worked 14–16 hours a day most days on the job, long after the others had gone home. Of course, good help is hard to find in a town built on tourism, but I did it, and the men were keen to work. When the Sizzler store was due to open, I worked around the clock for two days straight, and I can remember how relieved most of the other workers also were when they rolled the bundle of carpet out to be laid that afternoon. There were about six of us 'catching a kip' on the roll of carpet before it was laid whilst they laid the felt underlay. The Kentucky Fried Chicken outlet had already been opened just next door, and we all celebrated the opening of the Sizzler restaurant in a party that night. I remember standing in the queue, asleep on my feet after many days of hard work.

The boss was the evilest man you'd ever met for other reasons besides the following. I finished the restaurant party on Saturday night into the early hours of Sunday morning, and he rang me at 8 a.m. that Sunday and asked if I was ready to go to the next job, which was 3,400 km from Cairns. I told him straight, 'I wanted to take some time off to have a look around Cairns before I left for the next job.' He said that was fine with him.

Cathy and I went for a drive up into the mountains that day and planned a boat trip out onto the Great Barrier Reef the next day, followed by a drive sometime during the next week up into the Daintree Rainforest, but he rang me again that same day and said, 'Have you finished having a look around? You are needed at that next job.'

I spat the dummy (as you do) and threw all my shit into the van. I then left Cathy behind and drove 2,000 kilometres south straight back to Brisbane, picked up the apprentice the next morning, and headed off to the next job, which was over 1,400 kilometres further south away from my home town of Brisbane. I asked the apprentice to drive for a while whilst I tried to sleep in the back of the van, but I couldn't sleep; my adrenaline was pumping. I was manufacturing that natural body speed. I had driven twenty-two hours from Cairns and another sixteen hours to the next job south of Sydney. I stepped out of the van and slept for just two hours in a motel.

The boss was waiting at the motel where we stayed near the job. He wasn't too happy that I'd woken him from a great night sleep at 4 a.m. And at six, we headed out for the next job, which was another Sizzler restaurant. This job was also behind schedule; the first day was sixteen hours long, and we worked other days in some long hours to catch up. The end of this job was no different. It was around the clock for the last few days until it was completed.

Whilst working there over the next couple of weeks, it started. Cathy began ringing the motel where we stayed, accusing me of having played up the night before. I can honestly say that I barely had the energy to eat most nights, let alone play up. Where would I have found the time? The apprentice was also by my side twenty-four hours a day. She had rung me at the motel because there were no mobile phones in those days except for that huge brick by Motorola, which was just unaffordable for me during that time.

The money was good in that job, but the man-hours from my side of the table took their toll. I ended up having to take a long break after returning home, not being able to work at all in those days due to work overload. It was just before I left the company that we saw an incredibly cheap house for sale just across from my house on the same housing block we lived in. I was told about the house before it went onto the market and that it was extremely run-down, but it was a bargain I just couldn't pass up on.

In those days, a house around that same area was selling for $150,000, and they were asking just $77,000, but it was hardly what you would call a complete house. When you opened the door to the toilet, the first thing you saw was duct tape around the porcelain pan. It looked like someone had opened the door and kicked it in with a steel-capped boot, and the remainder of the house was no better.

It took a bit of convincing, but Cathy and I bought the house, going halves with her. In the relatively short span before buying the house, whilst we were together, she had saved up $12,000 whilst I sold my Harley-Davidson for $13,000 and used them as deposit. We didn't have any money left over for renovations, but we sent a backhoe and truck through the yard to clean up when we arrived. I had swapped some electrical work on the house that belonged to the backhoe driver for those few hours of machine work. He pulled two ten-cubic-metre truckloads of old furniture and rubbish out of the yard and house.

It was a mess, but what a bargain. Cathy was starting to see what I was talking about. There were actually people living in the house before we bought it. Now when I say 'living'—well, you know what I mean. They were there nonetheless.

One of those guys stayed on with us after we moved into the dump. He was a guy whom Cathy had gone to school with and was a talented welder, but after a while, due to his uncleanliness, he had to also go.

Cathy and I were sort of settling down, despite the ongoing arguments. Whenever Cathy would go out, she would drink a bottle of Scotch whilst out with friends, and that caused trouble in our relationship. She was forever starting to fight with me in front of my friends about subjects we'd been arguing about at home.

Cathy had brought another friend of hers from work (at Social Security) to live with us. She was only paying a very measly amount of rent for the room because it was hardly worth anything due to the run-down state of the house, and it was more of a nuisance for me because I was limited to the hours I could work on the house. It needed absolutely everything redone.

When we bought the house and moved in, I made one huge almost fatal mistake. I rented my house out to those Spanish people I had mentioned earlier, and they paid me in cash. I came up with this idea one night whilst on the grog with Cathy before we bought the house. I would rent out my house to someone at an irresistible cash price, and we would put the title deed of the run-down house into Cathy's name to avoid paying tax on the rental. What a brilliant idea. NOT.

I had started with the renovations to the old dump by painting the outside of the house. Cathy wasn't used to her boyfriend working a job and coming home to the house to work on the house. She wanted me to come home from work and place all my attention on her (I guess). She started putting on a turn in attention-seeking stunts, but I kept working as hard as I could, trying my best to ignore her antics. Looking back, that was my next major mistake. I guess I should have talked to her more during the attention seeking.

This was when the trouble truly began. Cathy and I were paid a visit by her future sister-in-law. Her name was Kaylene, and she was a heroin addict. Before Kaylene had arrived Cathy told me that she had set up her brother on an assault charge, telling the police that he had broken her jaw in an argument when it had actually been Kaylene who had broken her own jaw. I honestly couldn't believe what I was hearing. This girl had somehow broken her own jaw so as to set up Cathy's brother in an attempt to have him stay with her in the relationship. Kaylene was coming over to our house to celebrate the good news. She had told the police the truth about her jaw. She had told them that she had broken it herself. Cathy and I gave her a big pat on the back and asked her about the details of her jaw. I can't remember anything what she told us that day except that she had pumped her arm full of heroin to alleviate the pain when breaking her own jaw, but I must admit I wasn't really listening to what she had to say anyway.

Cathy and I had been living at the old dump for about twelve months when we both went to have a look at a big trail bike in the paper that was for sale and very cheap. I have always bought and sold cheap vehicles that I saw when I felt like it. When we went to the guy's place to look at the big trail bike, I had just finished telling him that the amount of money that I was making in an offer to him was every last cent I had to my name. He agreed to the price, and I bought the big trail bike for peanuts.

This fellow then said, 'All I have to do now is sell my Pontiac Trans Am that I have in the garage.' I couldn't resist the urge and asked him the price on the Trans Am. I honestly can't remember the price today, but I can't forget the way that it excited me even to this day as I write about it nearly thirty years later. I also can't remember what my excuse was for suddenly having that larger amount of money to offer, but I made him another lowballing offer on the car, and in his stressful financial situation, he couldn't resist.

I drove back home that afternoon and picked up around $12,000 cash for the car which I'd had stashed away somewhere. The guy was stoked when I returned with the cash, and Cathy was very cathartic about the great buy. We drove that car everywhere, and I felt like some dude who drove around selling drugs.

We needed some extra cash to help catch up on renovations and house payments when Cathy introduced me to some guy who had contacts in the security industry. Well, it was like security. He placed me on the front door of an illegal brothel to kick the butts of anyone who played up. There was never a dull moment on that door, and things usually went down in a hurry in that job, but it felt like it happened in slow motion whenever 'stuff happened'. I would drive my Trans Am to that place and park in the church just next door at around 10 p.m. for my shift. That joint would close at 3 a.m. or just before the sun came up, when I'd head back home for a snooze till it was time for everyone to get up, and we'd all go to work.

This door that I worked was at the edge of town and was up on a hill. How appropriate. It was (more or less) the house on the hill, but I wasn't there the night some drunk drove his Hillbilly ute into the front door and shut it down. Luckily for me, I was on a well-deserved night off.

That Hillbilly had brought way too much attention to the illegal dealings of that illegal brothel. I guess the police had to write a report on the events going on inside the building, and that was what shut them down, but it was good cash whilst it lasted. I got to do some exciting things when standing guard for the ladies and looking after the patrons. I never ever went near the girls.

After I worked that door, Cathy started to go a bit crazy with the drugs. She began revisiting friends whom she'd smoked drugs with before I met her and fell back into her old ways a bit. I kept working on the house, and I guess all the problems arose from those days on.

Cathy had taken the Trans Am to visit another sister-in-law and brother. When returning home, Cathy said on the phone just before she left that they'd smoked a bag of weed whilst on her visit. I didn't have any great attachment to the car and simply told her to take it easy on the way home.

When she arrived home, she came tearing into the house in one of her flusters and was jabbering something about having hit a lady with a baby in our street. I couldn't believe what I was hearing and jumped straight into my work van to get down there.

When I arrived, there was an Aboriginal woman lying on the footpath with a pillow under her head that someone had given her. There was a white lady standing above her, holding a little Aboriginal baby, and a few people standing around her as well. I asked if she was okay. They said an ambulance was on its way. Someone said that he was a doctor, and he said she was fine.

I stayed with her until the ambulance arrived and asked her what had happened. She said a car just simply drove round the corner and hit her, and then she flew up into the air. I returned home and told Cathy to have a stiff Scotch to calm down.

It wasn't long after that the police arrived at our home to interview Cathy and log the details of what had occurred. I stepped in and told the police that Cathy had just now been drinking Scotch to calm her down so as to cover her for drink-driving. If she had been drinking after the accident, then the police couldn't get a legal reading to charge her with drink-driving. I was trying to help Cathy, but she couldn't see that. Cathy thought I was trying to 'can her arse' for drink-driving, and she started to get all emotional and cathartic again.

The police never ever charged Cathy with leaving the scene of an accident or anything like that at all. To this day, I can never understand that. After the police left, Cathy started an argument with me for telling them she'd been drinking. Cathy would bring up that subject time and again in the future, and no matter how hard I tried, it seemed she could never understand.

All our real problems started from around that day. Someone suggested in the years to follow that maybe Cathy had been smoking skunk weed or weed that was grown in hydroponics. That stuff is supposed to put a lot of people in psychiatric hospitals with severe cases of paranoia.

It wasn't long after that Cathy spoke to me about a baby christening that she was going to. I guess I should have given the house renovations a big miss that day and gone with her to the christening instead. Cathy took the Trans Am again and bought another bottle of Scotch to drink. I never thought she'd ever mix drinking and driving together again, but she did.

I spent the day renovating the house whilst working on the stumps. It was about sundown when I heard the Trans Am come in the driveway. I'd finished working on the house for the day and had just sat down after cracking open my first beer for the day. Cathy charged in the door in her usual cathartic style and simply said, 'Which bedroom do you want?'

Man, this was news to me. I knew I wasn't in the good books after staying home to work on the house, but she was on fire with anger. (A full bottle of Scotch will do that.) I tried my very best not to enter into any argument and simply replied with 'The room with my bed in it, I guess.' You see, we had been using all my furniture in the house, so it seemed appropriate to me to sleep in my own bed.

Cathy went into our bedroom and proceeded to pull my bed to pieces so as to move it into the other spare room. There was a mighty ruckus going on in our bedroom, but from the pieces of bed that flew out of the bedroom and into the foyer, I had guessed she was pulling it apart. God only knew what she was going to sleep on. I couldn't see any reason for Cathy to be carrying on the way she was. I had done a good thing as far as I was concerned and had been fixing the stumps on the house. It really needed it.

She charged out of the bedroom and started putting up her fists, wanting a fist fight with me. I was really confused and just sat there in the chair, having an occasional sip on my beer, watching it all go by. I went to the fridge for another cold one when her catharsis really kicked in.

Cathy came into the lounge room with her fists up in the air, walked over to my much-loved coffee table, and flipped it over. It cracked, and a leg broke off. I very casually walked over to her favourite display of crystals (which was a small table full), picked up the whole tablecloth, and threw them all out the window. (Another wrong move, I know.) Then Cathy went to the back door and began yelling out to the neighbours, 'Mick Cox fucks his mum! Mick Cox has got herpes!'

I was really starting to wonder what I'd really done by now to deserve this. I really hadn't done anything besides working on the house to deserve this rubbish. I decided then and there to pack some clothes and get the hell out of that place. She could have her fat friend from work who was in her bedroom listening to everything, and I was outta there.

I put together a quick plan and decided to make a run for it to my mother's house until I could find a place to live. Who cared about the finances on the house? I just wanted out.

I went into the bedroom and opened the drawer to put some clothes into an overnight bag. Cathy asked me what I was doing. I said, 'I'm outta here. I've had enough of you.' The next thing I can remember I was being hit over the head with one of the timber slats from the bed base. I turned and took it off her after it had hurt my head.

Cathy started punching into me and took the bag of clothes away. I went for Cathy to place her on the floor whilst yelling at her to leave me alone. I held her on the floor with my elbow in her face, telling her to F off and to shut the F up.

I let her up off the floor. I hadn't hurt her in any way, and she then started throwing punches at me. I deflected them from my body, stepping back at the same time. She stood there saying more hurtful things at me again as I swung my right arm in a backhander across her face to ward her off, I had no intention of connecting with her. Just at that very same split second, two things occurred.

The big fat friend entered the room, and Cathy stepped forwards. I only just connected the back of my hand with her jaw, and the jaw went 'clunk', but it wasn't with any great force whatsoever, and it wouldn't have hurt her in any way. Then the cathartic Cathy, whom I knew only too well, started on at me. 'You've broken my jaw. You broke my jaw.'

I didn't think to reply to her with *If I've broken your jaw, then how come you can talk?*

The big fat chick left the room and went back to her TV in her room. She simply turned it up to drown out the arguing. Cathy went for the phone and rang the emergency number, I think, but I ripped the phone out of the wall and gently tossed it into her lap.

Cathy had now changed into my bathrobe and kept crying that I'd broken her jaw. I told her, 'If I'd broken your jaw, then you need to go to a hospital.' This was another brilliant plan I'd come up with, to take her to a hospital instead of just leaving. Had I left, she would have calmed down, and everything would have been all right unless, of course, she broke her own jaw after I left.

I took her by the robe, marched her down to the car, and placed her in the passenger seat. As I drove out of the yard, she was busy kicking the living hell out of the inside of my van, lying slumped in the seat. We hadn't gotten far down the road when she kicked the glass sunroof in and smashed it. Little squares of safety glass went everywhere, but I drove on.

I got to the end of our street, and by this time, she had calmed down just a fraction, so I decided to have a change of plans, turning right towards her sister's place instead of left to the hospital. I had decided to take her to her sister's house and to return home and pack some things—another bad idea.

Cathy tried her best to open the passenger door whilst we drove along, but I held on to her by the robe, stopping her. She calmed down just a fraction more, and her sister didn't live far from where we were. I arrived at her sister's street and then decided to let her out of the van just two doors away from the house and to then drive off as fast as I could, which was another major disastrous plan, and let me tell you why after a few more details.

That night, I went to my mother's house to stay for a while or indefinitely or until she calmed down when something very mysterious happened. I received a phone call the next day from Cathy's father saying that they had just picked her up (that day) from the hospital and that there were a few things that we all need to chat about. I had no idea. Her father, who was an extremely violent man, was very passive on the phone and simply said we all had to sit down and talk, so I drove back over to the house.

They were all downstairs underneath the old dungeon, waiting for me. Cathy's head was the size of a football; it was black and blue. The first thing I said when I saw her was 'I didn't do that.'

Her father said, 'Calm down. We all have to talk about this.'

I reiterated what I had just said in a feverous manner. 'I DIDN'T DO THAT!' It felt kind of strange, but nobody really responded to what I'd just said. Cathy was extremely quiet.

Her father just kept saying, 'You have to break the cycle.' I never really understood what he was on about, but I returned to the house, and everything was just about back to normal. Cathy's dad rang me up a few days later and told me that the police were on their way to our home to place some kind of order on me. He clicked into his violent mode and started threatening me about the order. He told me that if I didn't consent to the order being made, he was going to come and get me. (I clearly remember that as his exact words which scared the living shit out of me.) He was extremely violent when he went off. Cathy told me that he had broken her mother's arm one day in an argument.

The police placed a domestic violence order on me, and I tried my very best to stop it, but I was given the legal advice to just consent to it. Besides, her father was ringing and threatening me again and again.

Cathy took some time off work, and I stayed home and renovated. We never ever spoke about the night in question, and we lived together for another full year and a bit. It was around fourteen months later when the next step occurred, and it wasn't all good times whilst living together. Looking back, Cathy was forever seeking attention.

I had no idea what to say. She kept wanting to argue all the time about that damn car accident involving the young Aboriginal lady and her baby. Typically, Cathy would come home and say, 'I took cocaine today because of you.'

I would retort with 'You never took cocaine because of me. I never forced you into it.'

Cathy kept getting violent with me. One night she threw a big heavy spanner at me and hit me on the ankle, nearly breaking it. Another night she pointed a knife at me, and I grabbed it very carefully and snapped the blade off. I have no idea how that came off in my favour.

She crossed the line one day when I was under my work van, servicing the suspension. I had the van up on car ramps and was lying on the ground under the van when I heard some feet shuffling in the gravel above me. I swung my head out from under the van to see Cathy standing on top of the rock wall with a big boulder above her head, about to hit me with it. I swung back under the van, and the boulder hit the ground right where my head lay. I'd never forget those next few seconds lying under that van, saying to myself, *I gotta get the hell out of here, or I won't make it out alive.*

I was thinking during the following days that, between her and her father, I was going to be dead soon. I told her in the next week that I was going to my mother's to live and packed my clothes and got the hell outta there. I was living at my mother's house a few suburbs away, and Cathy tried her best to get me to move back home, but when I refused, she said that she was moving out. She went to her auntie's place and lived under her house at first. I moved back into the home because it needed more renovations.

Cathy got together with a young guy who came out with us on a few big booze-ups. This guy was from down south in Victoria, and he was a young guy who worked alongside me on installing a huge piece of machinery in a dog food factory. The kid was a welder, and Cathy was now together with him.

I have no idea to this day why I did it, but I went to visit Cathy—maybe because I was young and frisky or maybe through human instinct of jealousy. I don't know. Cathy wouldn't have anything to do with me at first, and so I left, but we went out together just one last time. It was full of friction.

It was about fifteen months after the incident that night, which was about a month after we split, while I was living in the house by myself that the police pulled me over on my motorcycle to tell me I was wanted for an outstanding arrest warrant. I went with them to the station and was charged with two accounts of grievous bodily harm. One was with the use of a weapon, namely, a telephone. I can tell you it wasn't a trunk call that I was charged with. (Sorry, I couldn't resist that one.)

It was all about the big night with Cathy. She had very obviously made a complaint of assault against me.

I sought legal advice again, and I was told that the officer couldn't do that because she only had twelve months to make an official complaint, but he had, and they charged me all the same. The officer who carried out the charge had told the judge at the committal hearing that he had started his investigation a week before the elapsed time, and the judge told the court that he accepted it under ruling ABC-123 of the law. The charge and warrant had been made a week after I'd sent my first solicitor's letter for settlement negotiations on the house.

I'll just make a personal note here that I'm not the brightest bulb in the pack. The police officer and my solicitor sat there together before the court committal hearing had begun and were chatting about their weekend. I had no idea who had actually charged me for the offence because the investigating officer wasn't there when I was charged.

My brother, who sat in the court next to me, said, 'The officer who charged you is sitting next to your solicitor, who is sitting right beside you?' I told my brother that he was delusional; there was no way that my solicitor would be sitting next to me, talking to the cop about their weekend who had me charged, but he was correct. In those days, I was as dumb as dog shit, and I sincerely mean that folks.

Anyhow, quite some time after the committal hearing, I went to trial, and my solicitor told me on the morning of the Trial that I had to roll with this. I asked what he meant by this, and he simply said, 'Plead guilty.'

I tried my best to argue it out with him, but he said there was no supporting evidence in my favour. The fat chick who lived with us was going to trial against me. I finally agreed, and I rolled, but it wasn't without voicing my opinion first.

During the trial, my solicitor said to the presiding judge that his client wished to plead guilty. I argued then and there with my solicitor under my breath, and the judge adjourned the court so that we could discuss the matter further. When we left to talk about my plea, that judge must have heard me screaming at my solicitor in the corridor, 'I didn't do anything to her! I'm not pleading guilty!' My solicitor tried his very best to calm me down and then threatened to walk out on me. I had no idea what to do, and so I went back in and pleaded guilty.

I'd never forget what that judge said to me after I put in my plea. I guess it was her final test on me before passing sentence on me when she said. 'I have never seen such a cowardly attack on a woman before in my life. Have you anything to say to me before I pass sentence on you?'

I replied with 'Your Honour, all I want is to get as far away from this woman as I can.'

The judge then said, 'I sentence you to 120 hours of community service and record no conviction.'

I thought, *Gee, that sounds all right. She must have known I was innocent.*

My solicitor told me that the charge was one level under attempted murder. This description isn't everything that occurred in the lead-up to the trial, and let me tell you now the most important incidental facts that truly changed my life.

After my solicitor sat there in the courtroom at the first trial, called the committal hearing, next to the officer who had me charged, I didn't feel so well. My mind began to race. I started having all these . . . well . . . deep-seated paranoid thoughts. Was this pair good buddies? Had my solicitor spoken to the officer on that weekend about my charge? Most importantly, where had this officer come from? He was always in a plain clothes.

All my answers came when I spoke to another woman from Cathy's office who was a member of her team. It was the mother of the christened baby. She came over to the house which was being renovated after Cathy moved out and told me that the officer was a plain-clothes detective who had worked in the same building as Cathy and her team. They all exchanged information about clients. Cathy would tell him if it was a drug dealer's house and vice versa.

This woman told me that Cathy had given the cop a blow job one Friday night in a private bar which was on top of the building where they all worked. She said the private bar was some kind of BYO set-up for all of the government employees in that building. The three of them were the only people in the bar late one Friday night. It was actually an unused caretakers' living quarters that they nicknamed Freddie's Flat.

The lady from the christening sat at my home, telling me all this just days before it went to trial, and said the sexual favour was in exchange for having me charged, but the lady with the information had refused to come to court. My guess is that they all had a threesome, and she couldn't afford having her husband know about the late-night drinks and fornication. I had told my solicitor about the incident in Freddie's Flat, but he said that didn't change the fact that I may have assaulted Cathy.

Just like the fourteen-month gap, Cathy had made her complaint or the fact that the complaint was made the week after I sent my first solicitor's letter for settlement on the house, the solicitor told me it was all irrelevant. That day, I sat there thinking this solicitor was a mighty good friend of the officer and appeared to be doing his best to cover him up. Maybe he too would have been found out for infidelity by 'his' wife.

The next thing that happened before the trial took quite some time to really sink in. It was the sister-in-law Kaylene who broke her own jaw. She contacted me and told me before the trial that she had some important information to sell me. She said that she needed $1,000 in exchange for the information. I refused to talk to Kaylene. After all, she was related in some way to Cathy, and I didn't want Cathy to know who my new girlfriend was at the time. I told Kaylene to F off and hung up on her. She rang me the next week and the week after that, but I did the same.

I actually decided to meet up with Kaylene, and she asked me to give her a lift on my motorbike to a house on the south side of the city. She told me that she and Cathy were the ones who had broken Cathy's face up that night. She told me this just as I was dropping her off at this house, and she also said, 'Cathy came to my house just after you dropped her off at her sister's house.'

Kaylene then asked me for a bucketload more cash to go to trial with me, but I had no intention of ever believing her. After all, the house she said I was driving her to was a dealer's house where she was picking up heroin. That was really why I left her there.

I tried in vain to contact Kaylene again before the trial, but it was futile. I became as frustrated as hell when I kept coming up empty. To continue working in my daily job and also on the house was driving me into the ground. I was beginning to lose my mind.

It was around this time that I began dating the woman of my dreams whose name was Kandy. I was head over heels in love. This woman was the most amazing person I had ever met in all my life. My whole world was coming to pieces, but I hadn't a care in the world after I met Kandy. That was exactly who and what she was.

By the time this had rolled by, I had my butt back in the seat of a Harley-Davidson, and most days Kandy and I would 'ride the wind' as free as a bird. We drank together like the world was ending just about every day, and my life seemed to be changing for the better, or so I thought.

The last time I spoke to the solicitor, I told him about the sister-in-law, and he said that he would try to postpone the trial until I found her.

Kandy and I were going out together for about twelve months when she took me aside for a heart-to-heart and told me that her actual boyfriend had returned from his travel, and they were both getting back together. Kandy told me then and there that they had an arrangement whilst he travelled away for twelve months. All this was news to me.

I couldn't believe my ears. Kandy told me then and many other times that she was crazy about me, but you know what? That's what candy does to you. It lifts you and then puts you in a low.

My mind was really coming to pieces now. I started to imagine all sorts of things and couldn't for the life of me find Kaylene. I was living in a house that was pulled completely apart, and so was my mind. I could no longer afford to renovate the house; it was at a stalemate.

I couldn't get my head around the situation. Cathy had set me up on a serious assault charge after her sister-in-law had pumped heroin into her arm and broken her face up. Cathy had done some sexual favours for the cop who had me charged fourteen months after the alleged incident. The only real evidence they had was the fat chick who really saw nothing, and now Kandy was going back to her boyfriend whom she hadn't really broken up with. Mass confusion for me.

My life was a mess. I was really beginning to lose the plot. I would walk down the street and begin hearing the thoughts of people I passed by. There were also other things I can't remember just now, but I do remember thinking, *I have to visit a shrink.* And so I did.

I sat with this top-notch psychologist, talking to him for an hour. When the session ran out, I turned to him and said, 'This might sound really weird to you, but do you know a good clairvoyant?'

He replied, 'Yes, I do, and you need it.'

———

He gave me the phone number, and I left his clinic with every intention to ring her. I did and rode down there to see her about what had been happening to my mind. It was the most logical answer I'd had for my mind. I was becoming clairvoyant.

I visited this lady who was and still is today a very well-known clairvoyant who wrote a column for a popular women's magazine. She lived close to where I live now these days. As soon as I sat down, I actually told this woman a number of things about herself, and she agreed with what I'd said. She pulled out these cards and laid them on the table. I'd never been to a clairvoyant, and she said that it wasn't really obvious.

This lady and I chatted like two old tradesmen about the spirit world and other tools of the trade that she used. After I left her house, if I hadn't begun to go crazy, then I was really about to. I really lost all control of my mind. She had told me to buy some crystals and cards and a silk cloth to keep them all in, and so I did.

Following her instructions, I had left the crystals outside in the full moon and had been doing all sorts of other really weird things before I said to myself, 'Hang on for just one goddam moment. All this just doesn't sit right with me.' I needed to stop everything that I'd been doing for just a moment and get back to basics.

I remember saying to myself, 'Now God is real, RIGHT? He had this Son named Jesus, RIGHT?' You see, I couldn't think past this story about Jesus. It was around this point that I decided to go and visit my own sister-in-law, Jan. She had a Bible and had gone to a Catholic school. So that was where I'd begin my journey.

I think I rang my sister-in-law to ask if she was home. Jan told me to come straight over, and that was when it got a little weirder by the minute. I was riding my Harley-Davidson into her street when I passed my brother in his car, going the opposite way. I flashed my high beam at him, and he tried to run me and my motorcycle off the road. No, I mean, he really tried to run me off the road when he came over to my side of the road.

I let it go. I went to their house to speak to his wife about God. My opening question to Jan was 'Who was this guy they called Jesus Christ?' She endeavoured to tell me exactly what she knew about Jesus, who was the Son of God. I stayed for one coffee and bid her farewell.

I was walking out to my Harley-Davidson, which was parked in her backyard, when I began thinking about other parts of history as I knew it. I was walking along with my helmet in my hand, thinking about the way that scientists dug up dinosaurs, which were millions of years old, and they can almost tell you what they ate for breakfast. And this guy Jesus, he existed only 2,000 years ago.

My mind was fixated on the facts, and I was thinking right then and there in that very moment how real it all is. I walked to my motorbike, thinking out aloud and saying, 'I have no idea where I am going in life, but, God, give me strength.'

It was just then when I put my helmet on my head and looked up to the sky that I saw a cross in the clouds. It wasn't any normal cross though. It had DOTS ON THE TIPS, and in the very second that I saw the cross, it all closed in and was gone. In that moment, I thought very strongly about God and could feel his presence all around me. Then I simply said, 'THANK YOU.'

I threw my leg over my bike and rode off into the night. That was a Friday night, and over the next two days and nights, I proceeded to have a nervous breakdown. Seeing that cross in the clouds was the straw that broke the camel's back for me. It was all way, way too much for me. I was no longer losing the plot. I had well and truly lost it.

It was Sunday when D-Day arrived for me. I was no longer on this planet but had decided to take a quick ride on my Harley-Davidson across town. This police utility had somehow snuck up behind me and was sounding his siren, telling me to pull over. I felt that everything was going to be okay until he walked off with my license. I'd lost my license due to an accumulation of points, and I was now headed for a bigger fine or maybe an arrest when he returned.

I decided not to wait. The police ute he was driving was a snail with a diesel motor compared with my Harley-Davidson, and it was then and there I decided to jump back on my motorbike and hightail it for the hills. Only God knows what was going through my mind or what I thought I was doing. He had my driver's license in his hand, a full view of my face in the open helmet, and the motorcycle make and colour.

I rode off at top speed through the traffic and the outer suburbs of Brisbane. I rode along a road that was about to be turned into a freeway which hadn't been repaired in years. I had the throttle handle screwed around to full throttle and was headed across this rough road at about 190 kilometres an hour. Did I mention that I rode absolutely everywhere with my Staffordshire bull terrier on the bike? At those speeds and rough terrain, the dog was just about flying off the side.

The bike was bouncing all over the road when I came up with the brilliant scheme to change the helmet I was wearing for the spare helmet that was hanging over the headlight. The helmet I was wearing was white, and the helmet on the headlight was red. It made great sense to me at the time to change my helmet to red because the cops would never recognise me with a dog sitting on the petrol tank. (I did say it seemed like a good idea to me at the time.)

This fancy manoeuvre began with taking off the white helmet and throwing it as far as I could off the road. I nearly came off with that ridiculous stunt when I reached over the dog to get the helmet off the headlight. I have no idea how I did it, but I did. I was now safe, or so I thought. I had changed to a red helmet, and the cops in front of me would never recognise me. Talk about an ostrich with its head in the sand.

I then literally headed for the hills to lie low for a while until the heat was off. I rode up to a small mountain on the outskirts of Brisbane called Clear Mountain. The fellow I had been working for had a house up there, and I thought I could have hidden out in his place during a surprise visit, but he wasn't home.

I sat there outside his house, wondering what to do, apparently not worried about my license the cop had taken from me or my license plate number. My mind was extremely stressed and all of a sudden, I realised that I had to be on the other side of Brisbane so as to rescue Kandy and the rest of the world from a fate worse than death, and I sincerely thought those exact words. The sudden realisation was followed by panic. How could I possibly get all the way over to the south side of Brisbane within a few minutes to save Kandy and the rest of the world? Then I realised it. I could fly.

I simply mounted the motorcycle, threw my dog on board, and headed out the driveway, which was about 100 metres long, at a million kilometres an hour. I screwed the throttle handle down so hard that I nearly snapped it, and by the time I reached the end of this driveway, I was doing about 120 kilometres an hour. Out of this driveway and across the other side of the road, there was a metal guardrail, followed by the great cliff.

Did I mention that, at this stage, I wasn't wearing a helmet? Luckily for me, I realised that I'd lost the plot but only about twenty metres before the guardrail as I went for every bit of the brakes I could find. At this point, I needed a ship anchor to stop and hadn't hit the brakes soon enough; but luckily for me, I T-boned the guardrail, doing about ninety kilometres per hour, but cannot remember anything after leaving the motorcycle seat. However, I do remember landing on the other side of the guardrail, right on the edge of the precipice. I looked over the edge and at my leg, all twisted up, thinking, *Oops, I've fucked up badly. The dog must have gone over that cliff.* My leg was at a ninety-degree angle to my knee in a sideways direction, and I grabbed my leg to straighten it out. My hip was also dislocated, but everything else was fine.

My motorcycle was on the other side of the guardrail, revving its guts out, and it had to be switched off before the motor was damaged, and so I crawled back over the boulders and rocks, dragging my dead limb behind me, and then clambered the guardrail. After turning off the bike, I laid there for a while, having some very conscious thoughts about what I had done, when the dog came back to me. He seemed fine; apparently, a fall over a steep cliff was all in a day's outing for him.

It was around this point when I flipped out again. I don't know how long it had been after that Evel Knievel stunt, but I then decided that I still had to move, and I began crawling down the road. Some guy who said he had heard my bike hit the rail was trying his very best to stop me from crawling down the road and doing more damage, but I swept his legs with my forearm and told him off. Luckily for me (and I mean it this time), he stopped me and held me on the other side of the road until the ambulance arrived. By this time, I had come back from never land and was lying there on the side of the road, chatting to this guy who stopped me from leaving the scene of the accident.

The ambulance driver took one look at me and asked, 'Have you always had one leg shorter than the other?'

I told him, 'NO, my hip is dislocated.'

To cut a long story short, I don't really remember the ride to the hospital, but I do remember flipping in and out of sanity. My thoughts were very lucid. I remember thinking at one stage that the ambulance driver was Satan and that I must go with him back to hell. I thought, *I had to cop it on the chin wherever I was destined for.*

In the hospital, the staff were great as usual. They rushed me in for an emergency operation to reinstate my dislocated hip after I had told them I hadn't eaten for days or even slept. (Well, you don't sleep when you're the son of God or whatever it was I had been thinking.) They put me under anaesthesia, and I was happy, of course. They were putting my hip back in place when I woke up from the anaesthetics. My leg was way up in the air and over my shoulder when I asked the guy holding it what he was up to. He told me, 'Nothing.' And he asked me to go back to sleep, and so I did. It was a good sleep too.

Whilst lying there in the hospital ward, I kept slipping in and out of reality; it was an extremely terrifying time. I kept having thoughts that I had to do this or that to save the world; otherwise, all humanity would suffer for an eternity. I didn't want to fail everyone, but I just wasn't on this planet during those moments; and looking back, I had only good intentions.

After a couple of days, they prepared me for an operation on my knee. I had smashed the cartilage, broken the kneecap into three pieces, and snapped three of the four ligaments in the knee joint. They opened up my knee and reinstated one of the ligaments using staples, drilled the kneecap down through the top, and pinned it back together using six-gauge fencing wire but surgical quality, of course. After the operation, I was put inside the psychiatric ward for three days of regulation, nothing too extensive.

They decided to put me back in the psychiatric ward for a while, and I was discharged from the general medical ward after about three months. This was about twenty-nine years ago today, and I remember most of it like it were five minutes ago. I also remember coming home with some girl out of the psychiatric ward who needed a place to stay for a while.

I only had one bedroom that was barely liveable in that house, which was pulled to pieces, and one bed that she and I slept in for the entire six months she stayed. I never touched the girl, and she was a great help to me. I couldn't drive or hobble very far.

I was down from work for a total of about nine months before I started running around the oval I lived next door to before getting ready to go back to work. I was working for about three months when I had a sudden relapse of the mind. I just couldn't stop thinking about that damn cross in the clouds. I ended up in the psychiatric ward again for another three days, but I was fine after only a few hours in there. I went back to work again, but let me tell you about the people I met along the way.

I was telling anyone who would listen about the cross in the clouds that I saw. I met this guy in the psychiatric ward who had been drunk and stoned during a car race across this old highway bridge in a nearby town. He had been racing across this long bridge and didn't make the turn at the other end, doing 100 miles per hour. He rolled the car into a petrol station, landing right between the petrol bowsers. He said that he was hanging upside down in the driver's seat when a guy in a long white robe opened his car door, undid his seatbelt, and helped him out of the car. He said he was standing there, wondering how it was possible he had survived without a scratch, when the cashier attendant came over to him. The attendant also said how lucky he was with all this fuel everywhere, and he said that it was all thanks to the tall guy in the white robe. The attendant asked him who it was he was talking about because he was the only person there on the scene. This guy in the hospital was all freaked out by the guy in the long white robe and explained to me that, the next thing he knew, the cops had brought him to the psychiatric ward.

Anytime I told my story of the cross, it was always to people I met who'd had similar experiences. And the more people I met with stories, the better I felt about my experience in the days and months that followed.

A little after the trial, I felt the urge to continue chasing Kaylene. I found out where her mother lived and couldn't believe my eyes. When I had dropped Cathy off near her sister's house, Kaylene lived just four doors away. I went to the house and spoke to Kaylene's mother. She couldn't remember a date but could clearly remember a night when Cathy had visited Kaylene around the time I had dropped her off in the car. She said that Cathy had never visited any other time.

I tried my very best to contact Kaylene, but it wasn't long after my visit with her mum that Kaylene hung herself on a rope at the clothes line at her mother's house. By this time, Cathy had also become a heroin addict, and it was all because Kaylene had pumped her arm full of junk before they broke Cathy's face that night, but I had no evidence about this evil pair and had hung up the phone on Kaylene so many times.

After I was back on my feet from hospital, I eventually met another girl and settled down a little bit. I never ever saw Kandy or Cathy ever again, but my supernatural experiences never stopped there. About every year, including the recent years, I would have a new experience. I'll tell you about some of the recent ones a little later in my short life story.

The new girlfriend had encouraged me to start drinking again, I guess through peer pressure rather than her prompting me. Drinking during the days of being spiritual isn't healthy for anyone, and I ended up in the hospital again a couple of more times before truly settling down. Looking back, my spirit guides were freaking me out so as to get me on the path. In fact, during the eighteen months after seeing the cross in the clouds, I ended up in the psychiatric ward three other times but was fine after that. It was those further apparitions or supernatural events that kept putting me there. As I just said, looking back, these events were bolstering my character into what I am today. Today I go to church every Sunday and sometimes twice. Some people need church to keep them on the path, but I always thoroughly enjoy meeting like minded new and old friends.

The next major step in my life was marriage. I had married a woman whom I felt was more my destiny rather than being in love with her. Let me explain that little statement. I met my wife through another lady, and the woman I married was related to all the people who were around me in my youth. I felt it was simply our destiny to be married.

I should have had a longer engagement so as to get to know her. It turned out that she wasn't a nice woman. She lied like hell and made my life worse than hell after we had a child. There was a good friend of her family who was in the police force which also didn't help me. He told my new mother-in-law that I had a police record after seriously assaulting my girlfriend (Cathy) of the time.

I had a child with this lying scoundrel, and I loved that child with all my heart, but that wasn't the outcome that she was happy with. From the first day that he spoke, I knew that I was in trouble. His first words were 'Dad, Dad, Dad'. You see, this woman lied her heart out to police, describing me as violent, just so that the police would evict me from the family home. Her aim was to cause as much hardship as possible so as to achieve full custody of the child. With full custody of the child, she would receive full financial support and of course be favoured in the family law court for the house.

Right now as I edit my book, I really struggle to write about this period in a short version. There is so much I'd like to say, but I will leave it there. Let me put it this way. It had been about ten years since I was in hospital before I fell out with my ex-wife, but she found plenty to complain about when we began arguing over contact with my son and the division of the finances. She made me out to be an absolute nutcase.

She lied like hell about me to police, who placed me under another order. She made up some horrific lies and pushed me to spend everything I had to just maintain a simple relationship with my son. I could never afford legal representation. In every direction I turned, she made trouble for me with lies, and maybe one day I'll tell you about them; they're worthy of their own book. Nothing was straightforward or simple, but I imagine there were people worse off than me.

During the years before meeting my ex-wife, I worked as an electrician during the week and worked on the weekends in the markets doing psychic readings, which I thoroughly enjoyed. At one stage of my life before marriage, psychic readings were my only source of income. I was a professional reader.

When I first returned to work after my initial breakdown, I was extremely sensitive and highly intuitive to those around me. I was working together with two other electricians, whose names were Trevor and Gus. All three of us often worked together as electricians and drank together on weekends, but after my breakdown, I had gone through a complete role reversal, never setting foot inside a pub when all three of us were working on the renovation of an old football club here in Brisbane. The youngest guy, Gus, was placed in charge of the job. It was (more or less) the first renovation he had been given charge of.

At the time during my recovery, I wasn't really in any condition to take control of the job, and we both guided Gus where he needed it, but he rarely needed any advice. He was very smart and had been taught by the best. Let me tell you about the experience that deeply involved Trevor.

Trevor

It was Easter weekend when we all broke up from work for the long weekend, with Good Friday being the following day and, as usual, a public holiday. Monday was also a public holiday, and Tuesday was our once-a-month construction industry rostered day off, which meant we had a five-day weekend ahead of us. Yahoo!

I went off to the markets to work as a reader for the long weekend, with church attendance on Sunday, whilst the two other electricians headed to the pub circuit. As I said above, I used to be the third musketeer on the weekends with these other two guys, but I had taken a different path to them after my first miracle and admission to hospital. I returned to work on Wednesday, meeting up with Gus on the construction site just before starting time.

Trevor was an older and a more diligent worker who normally arrived well before anyone else and opened up the site boxes. Trevor hadn't arrived yet, and I asked Gus where Trevor was. He had no idea, and so we started work. I said that it was very strange how Trevor hadn't rung to say he was going to be late. He'd never had a day off sick that I could ever recall and was always the first to arrive. Gus just shrugged and said, 'I don't know where he is.' We both worked until it was time for smoko, and Trevor still hadn't rung in to say he was sick or couldn't front up.

At smoko, I asked Gus what the pair of them had gotten up to on the long weekend, and he simply said they had drunk a few beers at one of the taverns. For Gus to say that they had a few beers, of course, meant they had gotten rotten drunk, and so I left it at that.

Just then as we stood up at the finish of smoko, we both turned to look out the huge window to see the boss arrive in his big four-wheel drive. We both walked outside and down the stairs to meet up with the boss as he walked up from the car park. Charlie said, 'How's it going, guys?' Charlie was the guy who owned the property up at Clear Mountain, where I had my accident on the Harley. (Accident?)

Gus and I greeted him and said hi. We both spoke to the boss about the job when, all of a sudden, Charlie asked where Trevor was. I said, 'He hasn't come in yet.' And I asked if he had rung him to say he was sick, to which Charlie replied no.

I then looked at Gus for a second time that day and said, 'What did you guys get up to on the weekend?'

Gus said, 'I don't know. The last thing I remember, there was a bit of a fight that we had at the tavern.'

Just then, as he said it, I got this queasy feeling in my stomach like I was about to be sick or at least close to it. Without an explanation, I then ran for my work van and drove off like a bat out of hell. As I drove past Gus and Charlie, they both said, 'Where are you going?'

I said, 'I'm headed for Trevor's house.' I drove like a madman to Trevor's house and arrived like Johnny-on-the-spot.

I charged straight up to Trevor's front door and knocked very loudly. A young guy I had never met opened the door, and I walked straight in. He said, 'Hey, who the fuck are you?'

I said, 'I'm Trevor's workmate, and he is supposed to be at work.'

He said, 'Trevor is in bed.' I asked which bedroom he lived in, and I went straight in.

When I opened the door, a stench hit me in the face. In his bedroom, there was a big pool of blood on the floor. It looked like he had thrown up blood, and I knew immediately that it was a sign of concussion. There were bloodied handprints all over the walls. It honestly looked like a murder scene. Trevor was asleep but wearing his sunglasses. I called to him, 'Trevor! Trevor!'

And he woke up, looking straight at me. He said, 'Coxy, what the HELL are you doing in my room?'

I said, 'It's Wednesday, Trevor, and you're supposed to be at work. You've been in a fight at the tavern, and it looks like you have concussion.'

He said, 'It's only Monday, and we don't go back to work until Wednesday.'

I said, 'It's Wednesday, Trevor.'

He told me to get the HELL out of his room, and I told him to come out into the kitchen, just outside his room. I was standing there, waiting as he came out, and he said, 'Tell me again why you were in my room.'

I told him that it was Wednesday, and he had been unconscious for a couple of days. He insisted that it was Monday, and I asked the young housemate how long Trevor had been asleep in his room. The kid said, 'He has been in there for three days.' And I queried the reason for not checking on Trevor.

I was just telling Trevor that he had to come to the hospital due to concussion when Charlie and Gus walked into the kitchen. Charlie asked what was going on, and I explained all the blood on the floor in his bedroom, and all the bloodied handprints on the wall. I said, 'He thinks it's Monday.'

Trevor insisted again that it actually was Monday, and I then begged him to come with me to the hospital. He told me that he was not going to any hospital and that he was fine. I told him that if he didn't come with me to the hospital, I was going to dial 000 (the emergency number). He told me to go right ahead. He said wasn't going to hospital.

I rang 000 and began describing the situation to them on the phone, asking them to send a car, when Trevor said, 'Okay, Coxy, hang up. I'll go with you to the hospital.' I hung up the phone, stuck him into my van, and drove like someone who was possessed again back into town to the hospital. We arrived at the hospital, and the guard on the gate asked 'what we wanted'. I told him that this fellow had concussion and needed the emergency, and he told us to park in the ambulance bay up the top.

We arrived in the ambulance bay, and I raced around to the passenger side of the van, expecting to help Trevor out of the van; but as he said, he was fine, and there was nothing obviously wrong with him. We both walked up to the emergency reception, and the nurse in charge asked what we wanted. I told her the brief story about Trevor, making sure to mention all the blood vomit and his memory loss, and she asked him his personal details for a check-over or suspected admission. Trevor told her his name, d.o.b. and address and assured her he was fine. She then instructed him to get into the wheelchair, and an orderly wheeled him into the emergency ward.

I sat in the waiting room, looking straight through to the bed where he laid. I was there for about ten minutes when there were suddenly squealing electronic noises and a commotion of medical staff around his bed. Trevor had died.

It may be a strange thing to say right now, but he had chosen the right place to do it. I went into the ward, and there were so many professional emergency staff around his bed that I could hardly see him. Luckily for everyone, they revived him and were keeping him stable when I was told to leave the emergency ward and to wait outside again. I was told that they took him into another part of the hospital where he would be under constant supervision.

I stayed for the remainder of the day and rang Charlie and Gus to tell them what was going on. Turned out Trevor had a blood clot on his brain, causing strokes. We all played the waiting game for the swelling to go down over two weeks in the intensive care unit. They told me during visits that Trevor died two other times whilst in intensive care over a two-month period, but they brought him back every time. His swelling on the brain went down, and they operated on him to remove the blood clots. After that, they inserted a metal plate in his head to replace the piece of skull which had been removed but there were serious side effects with the metal plate.

Along with losing that piece of his skull, he also lost the memory of his life. After he left high care, Trevor went into a psychiatric ward because he had inferior brain function and no memory for quite a while. I used to visit him in the ward because, out of all the people I knew, when I was in hospital, he was the only person who visited me regularly. Slowly, his memory came back to him in little snippets.

We had to tell him of his mother's death all over again, and he entered the grieving period all over again, but he said the greatest news he ever had was that he owned a Harley-Davidson which had an incredible paint job. I was told that they had to reintroduce him to his sister, and most of these memories made him happy.

He had about a year off work but eventually came back. The three musketeers were back together again, but it was short-lived.

Let me tell you why. The first job we all worked on was about to be electrically tested.

In my early days as an apprentice, I was indentured by the supply authority who used to test every installation; but today as electricians, we must test all the work we install. This job came up with faults on it. The two most vital wires, the red and green, had been reversed, and I had a strong feeling it was Trevor. Nobody to my knowledge had ever done that before. At first, we questioned the fact that Trevor always wore sunglasses, but he proved that it wasn't a problem.

In the next job we went to, I asked Trevor to fit off a row of power points. When he had finished, we tested them to find out how he went. The same problem had arisen, and that was when Trevor said he had become colour blind after the operation. I had no choice but to discretely tell Charlie about it because Trevor refused to.

Back in those days of the trade, it was not allowable to be colour blind in the electrical industry. Charlie said that it was regrettable, but he had no other work for Trevor to do other than electrical, and he had to lay Trevor off. It was only a small company.

Trevor went home and registered for unemployment benefit for the first time in his life, and it brought him down quite a bit when he was pensioned off. I felt guilt and told him one day that it was me who told Charlie about his colour blindness. I can't remember to this day how it was received, but I couldn't bear the guilt of being a snitch anymore. I used to visit Trevor most Friday afternoons after the week's work.

One day Trevor was in an incredibly happy mood on the phone before I arrived. He had gone out and bought himself a puppy. It was a long-haired German shepherd. It was a beautiful dog, and he gave it all the love in the world that he could give. He named the dog 'Magic' because he said that it was going to do magical things in the future.

He took Magic to puppy training school, to adolescent school, and then to adult training. He never missed a lesson and also taught Magic all sorts of wonderful things at home. When Magic was about 2 years old, Trevor applied for a job as a security guard in the train service for the state government. He and his best mate, Magic, had become security guards for the trains.

Trevor is still there today, and I can honestly say that it was the most magical job anyone could ever have, working nine to five with your best mate. I would have given my left arm to do that for a job.

The Supernatural Experiences

I mentioned earlier that I'd tell you about some of the supernatural and magical things that have gone on in my life starting with the night I saw the cross in the clouds. One of the most phenomenal incidents occurred at that job where Trevor, Gus, and I worked. There are an awful lot of things that have happened or I have witnessed throughout the years, especially after my mother passed away just four years ago in 2016. I'll tell you about a couple of those more recent ones first.

My mother bequeathed the home to me that I live in today. The short story is that I cared for her in the last 3½ years of her life, but there were other times during our life together that I helped out. I looked after my mother as much as I could throughout our entire life together but hands on for the last years. I had my own home, under a huge mortgage, just before she passed away and tried to rent it out whilst caring for her in the last years, but I had to sell it if I wasn't living in it. It was an older house that needed repair if it was to be rented out.

There were lots of supernatural events in my life, and I may write another book full of them, but an important supernatural event occurred on the first anniversary of her passing. I had the most endearing evidence that her spirit was around me. Before her anniversary, I had been praying fervently to my mother for a sign that she was well and happy in the spirit world. Before this, I'd had lots of communication in roundabout ways with her in the church. There were lots of overwhelming messages from mediums at church who hadn't ever met my mother that left me without any doubt she was there and looking after me.

This morning of her anniversary was incredible, and I warn you now to skip reading this if you have a weak heart. It is quite an impactive true story of mystical proportion. If you don't believe what I'm about to tell you, then you're not ready to hear the truth about the afterlife.

This anniversary day, I was hoping to see or hear some sort of sign from my mother. The reality of her passing was still very raw in my mind. That's how it is when you care for someone whilst they're ill in the final months before their passing. On the day of her anniversary, my partner had left early in the morning like she normally would. She spends most of her week caring for her daughter who has fibromyalgia and a day or two looking after her mother, who is about to hit 90 and lives across the state line. I often give them a helping hand by staying over and doing anything I can around the house, but this particular night, I had been blessed by my partner's presence on a night of staying over. I love her immensely today and always will.

Anyhow, I woke this morning at around five when she was leaving to go and look after her mum just across the state line, after which I went back to sleep. I eventually woke at around seven, knowing no one else was in the house, and wandered off to the toilet, which was upstairs on the second floor, where the bedrooms were of the town house. I came back to the bedroom and turned to face the wardrobe. When I was looking into it, trying to decide which shirt to wear for the day, it hit me like a brick to the back of my head. I slowly turned around and stood there, staring at my bed that had been made. I stood there looking at it, trying my best to remember the moments just a minute or so ago when I got out of bed. I was trying to remember making the bed.

I stood there thinking hard about the steps I made after getting out of bed and confirmed in my mind that I hadn't made the bed after getting out of it. It was an extremely weird emotion to be standing there, trying to remember the moments just a minute or two ago as if I was trying to convince myself that I had made the bed. I stood there looking at the bed when I realised it was extremely neat, and maybe it was my little lady who was still home who had made it. So I called out to her, but halfway through shouting her name, I realised that she had left early at five; and of course, there was no reply.

I walked around the bed in complete amazement at not only the fact that it was made but also how neatly it had been made. Then I heard my mother say to me, 'I was a steward on the overnight train, you know.' I just could not believe my eyes but also the loving and endearing energy which surrounded me in that moment. I had also forgotten for a moment that it was her anniversary.

I didn't freak out, although looking back today I know I should have. By the time this had happened to me in my life, I had grown accustomed to similar events and was very pleased that she was giving me such strong notice. I felt comforted by her sign and finally remembered it was her anniversary.

I eventually went downstairs for the day to start editing my book again when I rang my partner to tell her about the bed having been made. I was in an extremely jovial mood telling her about having woken up and gone to the toilet and then returning to grab a shirt, realising whilst standing there in the wardrobe that it was already made BY MY MUM.

My partner was very happy too, happy that I had received the sign that I'd wished for from my mother. I spent the remainder of the day downstairs editing and had used the downstairs bathroom whilst I was down there. I also made lunch and dinner whilst I was downstairs and had no real reason to venture back upstairs until it was bedtime again.

At nine thirty that night, I rang my partner again to tell her in a joking way that I was about to return to my bed for the night. 'I'm going back upstairs to go to the bed which was made by my mum.' She and I both had a great laugh. As I was walking up the stairs, I was telling her about my mother. When my mother, Betty, walked up these stairs every night, she would count the number of stairs until she reached the top. As a joke, I was counting the stairs as I reached the top. I turned to go into the bedroom, and there was my second miracle for the day.

The bed had been neatly turned back, waiting for me to get into it; and again, I felt a truly loving and endearing energy all around me, the type of energy you would feel when you were a kid and your mother kissed you on the forehead or something of that nature. Let me also say that there wasn't a thread of the sheets and blankets out of place.

I could tell you hundreds of stories like this, but this was the greatest event in the house ever since my mother passed away before, during, and after the renovations. My mother has always let me know she is there. Let me also say that it didn't happen only to me. I'll tell you about all my own supernatural experiences someday in another book.

There was this man who stayed with me for a few months. This guy is the most amazing person I think I will ever meet in my entire life. I'd had many spiritual experiences whilst with him to back up his statements. I'll tell you about his life in the next book, which will be about the spiritual and sometimes supernatural experiences that I've had throughout my entire life.

This guy is a paramount Navajo chief whom I'd met one day whilst doing psychic readings in the local markets. I was a medium in a psychic reading stall when he happened to walk past. It turns out that he was also a retired five-star admiral from the U.S. Navy. The incredible energy I felt before I met this guy was mind-blowing, but to hear his stories was truly an amazing time in my life.

His experience in the house with me was just another typical day in my life. I hadn't told him about my mother making the bed for me on that day, but he told me this amazing story whilst staying over. He would stay at my house whilst my partner was away, and when I asked him if he would need a blanket for the winter chills, he told me, 'No way.' He said that he never sleeps with any blankets or sheets on and that he leaves the window wide open at night during summer and winter. I told him that I had two quilts on my bed and that would be okay by me.

It was extremely funny one night in the middle of winter when I walked past his room with the door open. There he was, lying in the bed with the sheets and blankets pulled right up to his chin. He looked as snug as a bug in a rug there in that bed which belonged to my mother. He was nearly 90 years old when he stayed with me, and he kind of took on a parenting role.

In the morning, I asked him if he'd had a good night's sleep, and he said it was a total disaster. He said someone kept putting the blankets on me, and I kept throwing them onto the floor. I asked him what he meant by that, and he told me that he kept waking up with the blankets pulled up to his eyeballs. I then told him the story about my mother making my bed, especially how I returned to the bed that night only to arrive at a bed which was turned back. He said that was interesting and truly explained what he had experienced the night before.

The next time I got up during the night and walked past his bedroom, he was sleeping on top of the sheets and blankets. I guess he had stopped the spirits from covering him up with the blankets, but it didn't stop there. He said that he couldn't sleep one night and was sitting up, reading a magazine. He told me that he was truly enjoying leaning up against the wall, reading his magazine, and occasionally looking out to the night sky through the open window when suddenly the sliding glass window began to slowly close right before his very own eyes; and as if that wasn't enough, it was soon followed by the curtain closing right behind it. I told him nothing would surprise me in this house. My mother's presence is very strong.

Oh, they weren't the only things that ever happened in my mother's house. Lots of little things occurred. I've had my feet tickled, my toes pulled on several occasions, my hair stroked on one occasion and pulled on another, been poked in the ribs, my hand grabbed whilst hanging out of the bed and a glorious number of other little items. On one occasion I was in the bath half asleep and a spirit laid on-top of me. One fellow who rented a room said someone was knocking inside the wardrobe, once a toilet brush disappeared overnight, and a dustpan and brush disappeared from the middle of the garage floor during the day when it was there one minute and gone the next. My guess was that my mother was starting up a cleaning service in the spirit world.

I've had keys disappear from the table one day and reappear on the table the next day after turning the house upside down, looking for them, wanting to use the car. I had a huge spanner disappear from the floor one day whilst working on a machine and then reappear the same day, sitting in the middle of the floor. The list of events and incidents is long.

The first day meeting with the admiral was amazing. I had decided to go for a stroll with my partner through the local indoor markets that used to be in the shopping centre here in Beenleigh. We had just rounded a corner of the stalls when I saw this lady standing there, more or less shouting my name in astonishment. I sort of waved back to her and said, 'Hi, do I know you?'

She said that she had been to one of the churches when I was the guest for the Sunday night service. She said that she had always had this vision of Mick Cox doing readings in her psychic stall and hoped that one day it would come true. I told her that I had better go home, get my cards, and get my butt back there to make sure it came true.

She was truly ecstatic about that offer, and so I shortened my tour of the marketplace, headed home to have a shave, and put on a nice shirt. I was just driving out of my driveway when leaving home, and this female spirit spoke to me. She told me that she was sending her husband to see me in the marketplace, and she wanted me to tell him something. (To this day, I can't remember what the message was, but I do remember it was extremely personal.)

I agreed and drove down to the marketplace and set up my table. I've had my table for about twenty-five years, and it has a very ornate picture on it of an old man who is a seer. Getting out my table is like pulling on a set of steel-capped boots. By that, I mean that you know that it's time to go to work; but on this day, I didn't stay seated at my reading table.

I was standing near the entry door, keeping a keen eye out for the man that the lady in spirit had spoken to me about. The lady who owned the stall said to me, 'You're looking for someone, aren't you?' I told her that I was doing exactly that.

It was late that afternoon when this old guy on a set of crutches rounded the same corner of the marketplace. The spirit lady simply said, 'That's him.' And I waited for him to approach the stall.

He came by, and I said, 'Hello, sir, how are you doing today?'

He looked at me and retorted with 'Mighty fine, how are you?'

I told him, 'I am stoked to be here and of service to all the good folk like yourself.'

He looked at the psychic stall and asked what we were selling. I told him that we weren't really selling anything but were more or less offering a service. He said, 'Well, what's that?' I told him that we did clairvoyant readings and that it was a psychic stall.

He still had no idea what I was talking about, and so I said to him, 'We communicate with loved ones that have passed over.'

And he said, 'Is that a fact?'

I thought, *Well, now is as good a time as any to tell him.*

And so I said to him, 'You lost your wife a couple of years ago.'

He said, 'As a matter of a fact, I did.'

I said, 'Well, sir, she asked me to pass on a message to you. Would you like to know what it is?'

He looked at me with a big smirk on his face and an eerie disbelief and said, 'Sure, I would.'

I told him the message. (It was actually only about four words of an extremely private nature, like a nickname or something.) Well, we had to get him a chair. He couldn't stand up on those crutches anymore. He just kept saying over and again, 'Well, that's my wife.' He honestly went as white as a ghost.

When he calmed down, I asked him if he would like to have a reading for just $25. And he said that he would like that very much. When he sat down, I began by telling him about half a dozen names of people who were directly related to him. I began by telling him the name of his father and uncle and told him that they were here now, telling me their names. I told him personal messages from those people, and he was totally amazed.

I described to him about his uncle's stature who was inconceivably tall and asked him if he was an American Indian, to which he nodded and told me he was a paramount chief for the Navajo nation of America. I passed on personal messages from these loved ones in spirit and spoke to him about a very special meeting which had been organised with the Wasitues in America.

He couldn't help but be overwhelmed about everything, and he told me that the news about that meeting was very enlightening. He also told me that he was a retired five-star admiral and a few exciting things about his time in the U.S. Navy, but most of all, he listened to what I had to say from those in the spirit world. It was a truly wondrous reading.

I had finished the reading, he was very happy, and so was I. He asked me if we could meet for a coffee one day and chat some more. I thought, *Well, after all, there was no real effort made on my behalf. It was simply having a chat, but some of the participants to the conversation weren't present in the body.*

We met during that week for a coffee, and I started giving him names of buddies he had served with in the early days of the navy, names that weren't your average Joe like Broad Knife, Guns, and

Long Charlie, followed by endearing messages and descriptions of events that had occurred in their lives.

As I told you, the admiral would eventually come and stay with me during the week, and we both got on like we were long-lost brothers. In fact, the admiral asked me to become his 'blood brother', and I told him that I would like that very much. I became his blood brother, and eventually, he moved into my home with me, and we cooked together whilst having these little chats on the lounge. Amongst other things, he was a diesel fitter/engineer and an electrical engineer, so I showed him this book of theories, to which he was amazed.

The window and curtains that closed in front of him right before his eyes happened before he moved in, and there were days when I would give him twenty to thirty names and just as many messages from loved ones in the spirit world. One of those reading sessions went for about an hour and a half outside the house he was staying at before he moved in. He would hit me up on the phone and ask me to pick him up so that he could come over.

During this session outside the house where he lived I had given him about fifty names and messages from people who drank in the tavern he frequented back in Nashville, Tennessee. He never touched a drop in his life, but he played the guitar like he was possessed by the spirit of Jimi Hendrix. The admiral used to also play in that tavern. At nearly ninety he played as a back-up guitarist for a local band which played the pub circuit here in South East Queensland.

I can't tell you enough about the life of this amazing man. He began studying Buddhism at one stage of his earlier life and was asked by a Shaolin monk to enter the Shaolin temple where they taught him Buddhism along with martial arts where he would become a 'master of the arts'. He studied wing chun, the same style as Bruce Lee, and had the very same teacher who is known today as simply Ip Man. He fought in many armies and many battles. I actually told him the names of some Asian spirits who said they were Gurkhas, and he told me that he was a commander who led the Gurkhas on some missions. I'd channel the words of some of the spirit Gurkhas who reminded him of a story, and he would finish telling me the story where he was nothing short of heroic and amazing.

I would tell him names of people who had been with him throughout his life and sweet messages from them, and he would tell me awesome stories about his life. One day I told him a name of a man, and he said, 'No, I don't remember him.' I told him messages from the guy, but he still couldn't remember him. When I described a particular battle the spirit was describing where he had fought alongside this man, he finally remembered him in total.

I also told him one day the name Reg, and the admiral said, 'No, I spoke to Reg just recently, and he is alive and well.'

I said, 'Well, maybe this guy is a different bloke named Reg.' And I told him that Reg was an Interpol officer who was his superior. I told him a personal message from Reg, which brought the admiral to tears, and he said he couldn't believe he was gone. The very next day, the admiral's son rang him up to tell him that Reg had been killed, and the admiral told him that he already knew and also told him how he had known.

Over about a twelve-month period, I must have given the admiral over 300 names and personal messages from almost all of them; and just like all the others who stayed at my house, he too had received exactly what he needed from the 'house of healing'. You see, my mother told me to renovate the house in absolute white. She told me to lay white tiles on the floor throughout the house. She told me to paint the walls with the white ceiling paint and buy nothing but white furniture, and whoever stayed there would receive exactly what they needed to stay on the path. Most of the names I gave the admiral were given to him in the 'house of healing' except for those names from the tavern.

My mother brought a team of spirit workers into the house who call themselves Light. They aren't 'the' Light. They simply identify as Light. I've had people who needed to give up smoking come to stay, and they were personally helped by Light. I've had people who needed help with alcoholism or drugs, and they too were helped whilst at the house. I've had extremely lonely people visit the house, and they too regained the faith to trust in the spirit world when afterwards they suddenly met someone. I've even had a person who needed a car come to me for help in the house, and within minutes, an extremely cheap car had come online for sale that they were able to buy. The list of possibilities has been endless. Everyone who has come here has gotten something out of their experience, even if it is simply acceptance.

My mother told me, 'One day, people will come and stay here weekly for healing.' I pray that this day will come for all. I always look forward to hanging the white lantern out the front of the house when Light come to visit here again. The first day I bought the lantern, I had another spiritual experience whilst I was lying on the white lounge. The spirits entered the house and spoke to me about Ned Kelly, who was named as an outlaw in the late 1800s here in Australia. The entire Kelly family were there, visiting, and told me the following story of their lives. It was truly a love story.

Ned Kelly—The True Love Story

The story of Ned Kelly today told of a young man chased by the police who was told to 'stand and deliver' by a police officer when he was caught riding a stolen horse into town, but it wasn't the true course of events. Allow me to tell you what these spirits said about their life.

Ned Kelly was a young man whose passion lay in the sport of bare-knuckle boxing, and it only went to say that he was not only young but also a very virile man with good looks and a generous beard as well as heart. Let me especially tell you what the spirit of Ned Kelly told me about his heart and the romance that it found in a lady his age who was the rich grazier's daughter. It went a little like this.

Ned Kelly and the daughter of a rich grazier were very much in love with each other, but her father had placed a total ban on her ever seeing Ned. I'm sure that you'll agree that, when it comes to a rich grazier placing such strong boundaries on a precious daughter (especially in those days of the late 1800s), it also meant it was often followed by a boundary rule or a threat from the father, and he was no exception to this way of thinking.

His daughter was told 'to stay the hell away from Ned', and Ned was also warned by the man to stay far away from his daughter or else. It was the OR ELSE which met with Ned's demise. Ned and this young lady just couldn't stay away from each other, and that fact alone was what truly called for his arrest by the officer.

You see, Ned also had a wild side built into him by his father, who had already been arrested, prosecuted, and jailed for two years for stealing a jumbuck (a male sheep) to feed his family, and this was exactly the reason that the rich grazier had to stop Ned from seeing his daughter. You see, the rich grazier didn't want the son of a criminal to be seen with his daughter for all the reasons which went along with it, not that the grazier's threat was going to keep them apart, but their meetings had to be in secret, although the history today of Ned did actually speak very vaguely of his love for this woman.

As a wild young man, Ned's life was never dull. He'd often ride far up into the mountains and camp for days before returning. One day whilst up there in the mountains, he came across a bunch of horses that were tied up individually to trees, like they were in some sort of breeding scheme up there in a hideaway. They were now the property of a horse thief who had been rustling the horses.

This horse thief or bush ranger lived high in the mountains, where he corralled the horses and breed from the prize horses that he'd stolen. He would sometimes organise selling a newborn colt back to the very people he'd stolen the mare from. These stolen horses were hidden but also kept under a heavy and stealthy guard by the horse rustler.

At first, Ned would just watch this bush ranger to learn whatever secrets he could from him about horses, but one day he befriended the bush ranger and began to help him with the horses in exchange for learning how to be a horseman. On one particular day, he had gone to the mountains to work with the master thief when he noticed a beautiful palomino horse with a long golden mane that was unmistakably the horse which belonged to his secret love, the rich grazier's daughter.

Ned took advantage of the absence of the bush ranger, and without really thinking about the repercussions, he stole back the horse and was set to return it to the rightful owner. He had decided that, when returning the horse, he would tell the rich grazier that he simply found the horse in the foothills and had known it belonged to him, at the same time keeping quiet about knowing it belonged to his daughter. He didn't want to attract any attention. Ned knew that stealing the horse back from the bush ranger could cost him his life if he was caught.

He was successful in getting the horse from the bush ranger, who was a professional thief, but Ned was riding the well-known horse through town to get it back to the rich grazier when he was accosted by the officer, who yelled, 'THAT HORSE IS STOLEN! STAND AND DELIVER!'

The most unfortunate part of this whole story is that the officer had already drawn his gun on Ned before he had a chance to explain his story, and Ned had no choice but to follow a knee-jerk reaction which was to ride off on the horse. The officer had, of course, recognised Ned up there on the horse and went straight to the rich grazier to tell him that his horse had been found but, unfortunately, the horse thief was Ned Kelly, and he had gotten away with it again.

History doesn't tell us that Ned was trying to return the horse because Ned couldn't tell anyone about the love he shared with the rich grazier's daughter. This fact was actually the greatest miscarriage of justice ever carried out against Ned. History simply says that he was on a stolen horse, and it's true it was stolen, but he couldn't just tell the police that he knew where the horse was being held. That horse thief would have shot him dead after he was released from jail for horse rustling if he weren't hung.

When Ned had stolen the palomino horse back from the bush ranger, it was tied up, awaiting a stallion to be put over the horse for breeding. Ned had frayed the rope when he took the horse, making it look like the thin rope had been bitten through by the horse. The horse thief would never have known he stole it back, but the police officer in town had his gun drawn, giving Ned little chance. Luckily, he missed the target.

When the officer told him about Ned, the rich grazier gave explicit instructions to the police officer that if he shot Ned dead upon finding him, he would pay him a hefty sum of 500 pounds for making sure that Ned never saw his daughter ever again. The rich grazier had made this wicked promise to the police just outside the family homestead, knowing that his daughter was listening to what they were being promised. She knew where Ned was hiding and went there to warn Ned about what she had heard. The police knew that she would go there because that was the other cunning plan belonging to her father, who had simply followed a hunch that he had about his daughter and Ned Kelly. Ned was simply hiding in the make-shift campsite that he and his love had both shared so many times when making love under the moonlight.

You see, when those police officers went to arrest Ned, they had no intention of ever bringing him back alive. They were on a paid mission to assassinate Ned Kelly with the law behind them as judgement when he was trying to escape. Why was this police officer so vehement on making sure Ned was shot dead? The answer to that lay in a distressing short story about Ned's baby sister, Kate, and the police officer which makes it very obvious.

Ned had a sister named Kate Kelly. Our history today tells us that the police officer often visited Kate, who had only just become a teenager, and that he had placed Kate on his knee. That was, of course, just a polite way of telling us that he had raped her, but there was more to that than what history can tell us.

You see, when I myself were just 18 years old, I had a girlfriend whose mother belonged to a family who had adopted her at birth, and the grandmother (mother of the adopted woman) lived right next door to the Kellys in Beveridge, which was in the Victorian state of Australia, where this story took place. This lady (the adoptee mother) was visiting my girlfriend's family for Christmas dinner, where we all sat at the table. I sat opposite this old woman (who was about 92) when she suddenly touted the following words. 'Kate Kelly was raped.'

I guessed that either this old woman was becoming a bit senile or she was carrying information from her childhood which was quite overwhelming for the truth. I'm sure you'll agree that life sometimes gets like that. At this point, I had no idea who Kate Kelly even was, but I'm sure I was about to find out.

The adopted daughter (who was my girlfriend's mother) told her mother, 'Pipe down, Mum.'

After Christmas dinner together, I asked the old lady what she was on about, but her daughter just passed it off as some kind of joke, saying that her mother proclaimed to have known Ned Kelly's family, but she said it was impossible due to the years. After lunch, I waited for a moment when I was alone with the old woman from Victoria and I quietly asked her what she had meant about Kate Kelly and so she told me.

She said that her family were neighbours to the Kelly family in the small Victorian town and that her own father told her the story of the Kellys. 'Kate Kelly had been raped by that police officer.' I did the sums on this woman's age in the year that she told me (1990) and the year Ned Kelly was hung, 1880. It turned out that Ned and his gang had made history about twenty years before this woman was born, and that meant that the story was still very raw to those who knew the truth.

Ned Kelly was also renowned for his special attributes as a bare-knuckle boxer. The police also had a man who was a champion in the medieval sport. The Ned Kelly spirit told me that he had beaten the police in bare-knuckle tournaments behind the pub, and the police were never happy with the outcome of losing lots of money.

But there was more to my own indirect involvement with this Kelly, much, much more. You see, my father's grandfather was also a bare-knuckle boxer. And later here, I'll tell you that story and how it ties in with the Kellys.

So there was Ned and his brother, plus others, at the hideout when the three officers arrived. Ned knew that the officer had been promised a tidy sum for shooting him dead, which gave him no choice but to shoot to protect himself, even if it meant killing the police officer who had raped his sister, Kate. He had hoped that maybe then someone would listen to the Kellys about their sister's plight and about the rotten rich grazier, but it was to no avail. Ned had no proof, and he really couldn't tell everyone that he was having an affair. Back in those days, the discipline by a particular father was sometimes ruthless.

The Kellys were on the run. They needed guns and ammunition to survive, and that would cost them a lot of money. They were angered by the authorities and became rebellious. It was almost eighteen months since they had shot the officers.

The Queen hired a police officer from South Africa to come to Australia and hunt down the Kelly gang, and the story behind that man is second to none. Let me tell you about that South African man. He was hired due to his reputation to bring down the rascals who wreaked havoc on the community of South Africa.

Recently in 2010, I was talking to a man here in my neighbourhood in Beenleigh who came from South Africa. The rascals had broken into his home and had tampered with his wife. This guy told me that it was common in South Africa to carry a weapon. This would lower your chances of being attacked by the rascals. It was also an accepted or unrecorded practice in South Africa that if you shot dead a rascal in your own defence against an attack that you could take an ear of the rascal as a trophy.

The guy who told me this in Australia in 2010 was a kind-hearted man with a very placid nature. He and his family had gotten out of South Africa before it happened again and before his mind could become vengeful, but others he knew had applied the rule and had ears as trophies. It was a barbaric but true story.

Now the officer that the Queen had hired in 1880 to hunt down the Kelly gang lived in a South Africa, which was much more violent in those days and accepting of such rituals. Let me tell you what the spirit of Ned Kelly told me about that South African headhunter. He said, 'This man had trophy heads of every animal in the huge mansion he lived in back in South Africa.' He would often brag to the Australian officers about his collection, which included the heads of some Zulu warriors who had become what they deemed as a 'nuisance'. He kept these human trophies in his private room, which was his office in the home. They were there just to remind him of his special victories. The Ned Kelly spirit told me he was a genuine headhunter. If you don't think that such people lived in those days, then you are very naive. It was in 1880, over 140 years ago.

So this man was hired by the English monarchs to look after their little convict country, and the South African officer wasn't here in Australia very long before the showdown of the Kelly gang began. I believe, from what I was told by the spirits of this family, that a part of the deal struck with this man was for him to take home the trophy of Ned Kelly's head to hang it in his room, alongside those of Zulu warriors. He had only been in Australia about a week before he accosted the Kelly gang. He was obviously paid a lot of money and wanted another trophy head for his room, besides heightening his reputation.

Ned Kelly and his gang of thieves were held up in the Glenrowan Hotel as history tells us. Now let me tell you what the Kelly spirits told me about the Glenrowan siege and about the years that followed. After the South African officer (or barbaric headhunter) had Ned Kelly in his possession at the Glenrowan Hotel, he no longer needed the other members of the Kelly gang, and so he torched the hotel, knowing there were also civilians in there. The officer had been told that all those who remained inside had been drinking with the Kelly gang and rejoicing in the folly. He had no use for the heads of those who remained in the hotel, and so he torched every one of them. He saw them all as wretched souls, but that wasn't where the story ended for some of those men.

Ned Kelly came out of the hotel before the fire and was shot in the knees by an officer with a shotgun. His spirit told me that he lay on the ground outside the courthouse where some of his last photos were taken when the South African officer stood on his knee whilst asking him who remained in the Glenrowan hotel and told him not to try and escape, or there would be hell to pay.

I was also told by the spirit of his brother, Dan Kelly who was also in the hotel, that the police continually fired on the remaining people, and it was either stay there in the hotel and end your life quickly or be tortured by the police who were going to shoot you as well. After all, those in the hotel were watching what had happened to Ned outside. The spirit of Dan Kelly told me that he had escaped the fire, as well as Steve Hart. He said that they both hid in the cellar of the hotel, and recently in 2001, a private organisation scanned the original site of the fire and found traces of a very small cellar underground. The cellar scan revealed that it was about one metre square. Maybe it was where they held the monies for safekeeping as well. Dan Kelly and Steve Hart escaped from the fire in Victoria and ran to the state in the north (NSW), where they worked on a farm for two years. Let me bring you up to date with that story.

In the 1900s, a man came forward in Toowoomba and announced to his family on his deathbed that he was Dan Kelly, the brother of Ned Kelly. He told his family around the bedside that he had escaped from the Glenrowan fire along with Steve Hart when they both hid in the cellar. This man had bad burns scarring that was well known to the family. He eventually said it was caused by the fire at Glenrowan.

He said that they both escaped and ran to NSW, where they worked on a farm for two years. After the story was told by this man, they found the farm and proved that the two men had, in fact, worked there for two years around the time of the Kelly gang. They both then moved from NSW onto Toowoomba in Queensland.

Dan Kelly stayed in Toowoomba, but Steve Hart moved on to live the remainder of his life near Gympie in the south-eastern part of Queensland. If Ned Kelly was a bare-knuckle boxer, then it only stood to reason that Dan Kelly was also gifted. One day recently in 2010 my own father showed me a front-page newspaper clipping that showed this very same man in a tournament fight in the early part of the 1900s against his grandfather in Toowoomba. The man proclaiming to be Dan Kelly won that fight. The newspaper clipping showed the two boxers in a stance, facing each other before the tournament. All my family were avid boxers, including during the time I spent there in my youth.

I am not completely certain, but I think the woman that Dan Kelly raised a family with in Toowoomba was an Aboriginal lady, and these people today in Australia are still limited to what they can and can't say. There is another twist to this story. In 2001, they made the movie about The Kelly Gang with actors Orlando Bloom and the Australian, Heath Ledger. The two actors went straight to Toowoomba before making the movie and spent two months there. I bet it was to talk to the entire family who were related to this man.

Dan Kelly lived out the remainder of his life in Toowoomba whilst Kate Kelly went through a living hell and back. History tells us that Kate Kelly went quite mad after her brother was hung in the Melbourne jail. People told the story that Kate was always on the move from town to town, telling weird stories about her brothers and the authorities.

The spirit of Kate Kelly told me that after Ned was hung by the authorities, the butchers took his head, stripped it to the skull, and put on display in the Melbourne jail, or so they said. She told me it wasn't his head on display. She said that Ned had different features. This was the crazy story she was telling everyone. The spirit told me that she had gone to the jail to look at the skull, but she reportedly began telling people that it wasn't Ned's head on display; it didn't have any of his features.

You see, Kate was also Faye. This was the nickname given to those who spoke to the spirits. Today we call them mediums. Kate was Faye and had been told by the spirit of her brother at the time that it wasn't his skull on display. All this got her a reputation of being crazy.

History today also tells us that Kate was continually on the move from town to town, telling weird stories about Ned. (Well, you can imagine why they felt this way back in those days.) Kate today is described as being very unstable, and here is the reason why. The spirit of Kate told me that she received a warning from the authorities to keep quiet about the skull, but she refused to heed the warning. It was said that Kate continually moved from one property to the next, never settling down, but she wanted to tell as many people as she could who supported Ned.

History actually also tells us that Kate Kelly drowned in twelve inches of water where it was reported that she had committed suicide, but we can all see that this wasn't the case. Someone had been looking for Kate as she travelled from town to town, forever changing her name. At the time, Kate also had the story to tell about the police officer who had raped her.

You may also be able to understand that the barbaric officer who took Ned's head back to South Africa was under the authority of the monarch, and they couldn't afford to have such a barbaric story being told about a headhunter in association with the royal name. The spirit of Kate Kelly also told me that she had to pay a lot of money for ammunition and better guns that she supplied to the boys whilst they were on the run. The authorities were also finding all this out about Kate as time went by.

It was around 2001, at the same time that the two actors came out to Australia, that they discovered that the skull on display wasn't Ned's. Heath Ledger, who spoke to the family of Dan Kelly, was told the true story. Maybe that family lives in fear of repercussion today. You can only imagine the persecutions that family has already been through for being Aboriginal along with the stolen generation. The spirit of Heath Ledger told me about the skull, and he also told me that you can see his anger that he purged out in his acting when he made the *Batman* movie as the Joker before he died.

When the authorities were told about the skull on display in the Melbourne jail, they decided to invest in a search for the true skull. The body of Ned Kelly, less his head which was supposed to be on display for the past 140yrs in the Melbourne jail, was buried in a mass grave at the site of the old Melbourne jail, and so they dug up the bodies in the mass grave. They found the bones of the body which had been shot in the legs by a shotgun. I believe they may have also carried out a DNA test on the bones to match them up with a teacher who lives in Victoria today, said to be related to the Kellys.

One thing is for sure. In 2001, they never found a head which belonged to Ned Kelly in that mass grave in Australia, but maybe they should have looked in a rubbish dump somewhere in South Africa after the family of the headhunter threw it there. History today tells us that there were actually weird rumours of Ned Kelly having been seen on the battlefields of the South African Boer War, but I believe that these were just contorted stories that Kate told everyone about his head having gone to South Africa.

They were the last words the Kelly spirits whispered to me on the lounge in the house of healing.

The other night (in 2020) whilst asleep in my bed, I was woken by the spirit of Ned Kelly, who told me that there is a tree just ten miles from the family home which has his initials carved into it. It was only possible to hear the Kelly spirit for a few seconds, and so I'm not exactly certain if he said it also has the initials of his lover.